At War Within

At War Within

The Double-Edged Sword of Immunity

William R. Clark

Oxford University Press
New York   Oxford

Oxford University Press

Oxford   New York
Athens   Auckland   Bangkok   Bogota   Bombay   Buenos Aires
Calcutta   Cape Town   Dar es Salaam   Delhi   Florence   Hong Kong
Istanbul   Karachi   Kuala Lumpur   Madras   Madrid   Melbourne
Mexico City   Nairobi   Paris   Singapore   Taipei   Tokyo   Toronto

and associated companies in
Berlin   Ibadan

Copyright © 1995 by Oxford University Press, Inc.

First published by Oxford University Press, Inc., 1995
First issued as an Oxford University Press paperback, 1997

Oxford is a registered trademark of Oxford University Press

Library of Congress Cataloging-in-Publication Data
Clark, William R., 1938–
At War Within : the double-edged sword
of immunity / William R. Clark.
p. cm.   Includes bibliographical references and index.
ISBN 0-19-509286-4 (acid-free paper)
ISBN 0–19-511568–6 (Pbk.)
1. Immune system—Popular works.
2. Immunologic diseases—
Popular works.   I. Title.
QR181.7.C624   1995
616.07'9—dc20   94-45134

1 3 5 7 9 10 8 6 4 2

Printed in the United States of America

Contents

Introduction

Most of us are aware that the immune system is designed to protect us from the thousands upon thousands of predatory microorganisms that can invade and seriously damage virtually every part of the body. Millions of years of evolution have honed it to do just that, in animals and in humans. Our immune systems are finely tuned, highly integrated defense complexes that relentlessly track, identify, and destroy a wide range of would-be body crashers. Once the immune system has an unwanted foreign invader in its sights, it can bring a formidable array of chemical and cellular weapons to bear on its elimination. This is the side of the immune system of which most of us are aware—the nurturing and protective side.

The immune system can and does provide a powerful defense against potential pathogens, but what is perhaps less obvious is that it is also capable of bringing *too much* power to bear during the course of clearing away foreign invaders. Like an army lashing out blindly against an unseen and unmeasured enemy, the immune system is capable of using excessive deadly force in the wrong time or place—and it is capable of overkill. And as almost always happens in such situations, the most devastating damage of all may be done to innocent bystanders. The result could be nothing more than a mildly annoying allergy. But it can be more deadly. People may die from hepatitis, not because the virus destroys the liver—the virus itself is actually quite harmless—but because of the violence of the attack of the immune system on the infected liver. The same may be true for the lung damage seen in tuberculosis. A great deal of the degeneration of our bodies as we grow older may be due to subclinical autoimmune disease.

The immune system bedevils us in other ways. The immune system is *the* major barrier to organ transplantation. Healthy donor organs that could save the lives of individuals suffering from end-stage heart or kidney disease are violently rejected by the immune system. Bone marrow transplants that could save the lives of leukemia victims or children dying of immune deficiency diseases fail because of immunological complications. Yet the immune system apparently fails to protect us in the case of most of the cancers that afflict us. In AIDS, the loss of immune function that is the hallmark of this disease may be due as much to the immune system attacking itself as to damage from the AIDS virus.

Why do these things happen? We can never know for sure. Part of the problem may well be that we humans, uniquely among the creatures of the earth, have managed to stay alive fifty or sixty years beyond our prime breeding years. Nature never expected that. Our immune systems were designed to keep us alive only long enough to reproduce. As we have extended our life span through science and technology, we have increasingly become the victims of the cumulative effects of the lethal efficiency—and sometimes the bumbling overzealousness—of our own immune systems.

Yet as we know only too well from immune deficiency diseases, if the immune system stops working, we can die in a matter of days or weeks, unless we want to live out our lives in a sterile bubble. The immune system acts very much like a sixth sense, helping the brain to detect the presence of potentially deadly pathogens within us, and mobilizing the body to deal with them. In human beings, at least, the loss of any of our five primary senses can be managed, but the loss of our immune systems is, without some sort of intervention, uniformly fatal.

Our immune systems are thus like a high-wire balancing act. Death lies on either side. Science and medicine have given us the means to keep our balance for most of the length of the wire, but it is still a very risky act.

At War Within

ONE

Overture to a Science Unborn:
Smallpox and the Origins
of Immunology

What does it really mean to be immune to something? To a citizen of ancient Rome (where the word originated) it meant freedom from some onerous duty owed to the state. A latter-day example of this came during the American Civil War, when draftees could pay money to the government to be excused from military service—they were in effect buying "immunity" from their obligation to take a chance on being killed. The medical sense in which we now use this term is a fairly recent adaptation. "Being immune" to something means that, as a result of having once been exposed to something our bodies consider foreign, the immune system reacts more strongly to it the second time around. This is a special property of the immune system called *memory*. Among all the marvelous and sophisticated parts of the body, only the immune system and the brain share the property of memory—the ability to store information about previous experience. This is a coincidence that both immunologists and neurobiologists find fascinating, and which many of them are actively exploring.

Although the present use of the term *immunity* is a relatively recent development, human consciousness of biological immunity seems to be very old. In fact, it is so old we don't really know when it first crept into human consciousness. As a concept, it evolved to describe what was apparently a fairly common observation—if someone came down with a particular disease, and somehow managed to survive it, he or she seldom ever got that disease again. One of the earliest references to this concept of immunity occurs in a description of the plague that devastated Athens in 430 B.C. This description was written by a minor Athenian general named Thucydides, in his amazingly comprehensive *History of the Peloponnesian War* (between Athens and Sparta in the fifth century B.C.). Thucydides, who had himself been stricken with and recovered from the plague, made a number of interesting observations concerning the effect of this disease on individuals and on society at large in stricken Athens. Some of his more fascinating observations concerned the almost total moral, social, political, and economic decay generated among an otherwise civilized people who came to believe there could very well be no tomorrow. Buried among these are several observations on the medical implications of the plague. (The precise nature of the disease represented by the Athens plague has never been entirely clear. It seems unlikely to have been related to the plagues that struck Europe in the Middle Ages.)

Here are a few of the more clinically relevant observations passed on to us by Thucydides, as translated by Rex Warner:*

> At the beginning, the doctors were quite incapable of treating the disease becuase of their ignorance of the right methods. In fact mortality among the doctors was the highest of all, since they came more frequently in contact with the sick.

Perhaps not a very charitable view of what must have been a tough pull for the poor (but obviously dedicated) physicians of the time. Thucydides then touches upon a theme that has fascinated immunologists for some time, although it has only very recently become a subject of rigorous scientific inquiry: the possible connection between psychological states and immunological resistance to disease:

*Quotations from *The Peloponnesian War* by Thucydides, translated by Rex Warner (Penguin Classics, 1954) copyright © Rex Warner, 1954, p. 154. Reproduced by permission of Penguin Books Ltd.

The most terrible thing of all was the despair into which people fell when they realized that they had caught the plague; for they would immediately adopt an attitude of utter hopelessness, and, by giving in this way, would lose their powers of resistance.

We shall have more to say later on about the interaction of the immune system with the mind. But now let us look at the most telling statement of all, which shows that the general concept of immunity was clearly recognized nearly twenty-five hundred years ago:

> Yet still the ones who felt most pity for the sick and the dying were those who had had the plague themselves and recovered from it. They knew what it was like and at the same time felt themselves to be safe, for *no one caught the desease twice, or if he did, the second attack was never fatal.* (Italics added.)

This definition of immunity to a specific disease is as valid today as it was two and a half millennia ago. Similar descriptions appeared from time to time in the ensuing centuries—for example, in Procopius's description of another plague that broke out in Constantinople in A.D. 542. Whether the pattern of "exposure–recovery–immunity" was recognized for diseases other then the plague we cannot be sure; if it was, no one ever wrote it down. People got sick all the time, without really knowing what was happening to them. The notion of discrete, identifiable diseases, each one with a distinct cause, is a relatively recent development in medical science. But throughout history there have always been disorders that, because of the reproducible, outwardly visible effects they had on the human body, could be recognized as distinct diseases, and in such cases one imagines that the exposure–recovery–immunity cycle may well have been recognized quite early in human history. One disease that readily fits into this category is smallpox.

No one knows exactly when or where in the world smallpox originated. There are pox-like diseases in domesticated animals, and it is possible that humans contracted it ten thousand or so years ago when they first began to maintain herds of livestock.

The mummified face, neck, and shoulders of Ramses V of Egypt, who died in the twelfth century B.C. from a "sudden illness," looks very much like that of a victim of one of the more virulent types of smallpox. Although smallpox was generally thought to have existed for thousands of years in China and India, the first written description of a disease that is almost certainly smallpox does not appear until the tenth century in a work by the Persian physician Rhazes, who lived and worked in Baghdad. Rhazes makes no mention of people who survived smallpox as being more resistant to it, although they certainly would have been.

Smallpox apparently was known in Greece or Rome—it may have killed Marcus Aurelius—but does not seem to have been a major health threat. It is not described in the numerous classical medical treatises that have come down to us from that period. There are also no Greek or Latin word roots relating to smallpox, another sign that the disease was not generally known in these cultures. On the other hand, Rhazes felt that almost all members of his society would likely experience the disease at some time in their lives. Smallpox is thought to have entered Western Europe either via the Moorish invasions, or later with crusaders returning from the Levant. A particularly deadly form of smallpox was brought to the New World by European explorers, where it destroyed untold numbers of indigenous peoples.

It is difficult now to imagine that smallpox was once one of the deadliest diseases on this planet, probably equaling the plague in the total number of people killed throughout history. It has been estimated that during the seventeenth and eighteenth centuries in Europe four hundred thousand people died each year from this disease. Smallpox no longer even exists in the world; it was systematically hunted down and exterminated by a program of mass immunization coordinated by the World Health Organization after World War II. The last major outbreak in the United States was in New York in 1947, and the last confirmed case of smallpox in the United States was diagnosed in 1949. The last known case worldwide was in Somalia in 1977. The likelihood of a natural occurrence of smallpox is now considered so small that infants are no longer immunized against it. The possibility of complications from the vaccine are considered to outweigh any possible protective effects.

But in urban centers of seventeenth-century Europe it was a very different story. In cities like London it was assumed, just as it had been in Baghdad, that *everyone* would come down with smallpox at some point in his or her life, and the mortality rate among the affected could be as high as 25 percent. People became fatalistic about it, and simply accepted it as a part of life. By the seventeenth century it was also definitely recognized that revovery from smallpox did confer resistance to getting the disease again. Because it was observed that smallpox was hardest on the sick or undernourished, people would often deliberately expose their children to it, on the assumption that it was better to deal with it while one was young and healthy. As a result, among poor people in particular where this was practiced, smallpox was mostly a disease of the young. Those children who survived the exposure were protected to a considerable extent as adults. Those who didn't meant one less mouth to feed. So this process of deliberate exposure of the young, to the extent that it was practiced, was the nearest that people in the West came to a program for gaining some control over smallpox. As we shall see, people in the Middle East and Far East, perhaps having had a longer experience with the disease, developed a much more effective approach not only to controlling smallpox, but to preventing it.

Before getting to that, let us take a brief look at the disease itself, since few people today have any familiarity with it. Smallpox is caused by a virus called the *variola* virus, often referred to simply as the "poxvirus." Although it may have originated in animals, the form of variola that causes smallpox, like the viruses that cause polio and measles, affects only humans. The initial symptoms caused by variola are not unlike flu: fever, headache, and general achiness (particularly in the joints), and a tired feeling. But the fever does not break and can in fact get high enough to be damaging in its own right. After about a week a rash breaks out, which within a day or two begins to develop into blisters (*pustules*) on the skin, particularly around the face. It was not at all uncommon for these pustules to cover a victim's entire body, with small adjacent pustules merging together to form giant blisters. After a day or two the pustules burst, pouring out liquid pus and eventually forming scabs. If the victim survived, the scabs fell off, leaving behind a disfiguring, depigmented scar or "pock." The virus also wreaks havoc inside the body, causing extensive internal bleeding and the black vomit so often noted by early physicians.

Throughout its history on earth, smallpox, once it set in, was—like any viral disease—essentially untreatable. The more virulent strains could kill as many as 40 percent of the people they infected. All one could do was isolate the victim, make him or her as comfortable as possible, and let nature take its course. Even the advent of antibiotics about the time of World War II made no difference, because while antibiotics are very effective against bacteria, they have no effect on viruses.

As mentioned earlier, there is no written historical record indicating just *when* people recognized that exposure to smallpox would protect survivors from future attack by the disease, but it is difficult to imagine that this was not generally known for many centuries. The first reports of a means for conferring protection against smallpox *without going through a natural course of infection* began filtering into the West from Greece and Turkey in the early 1700s. Probably following a procedure developed earlier in China, the practice had evolved in the Middle East of exposing individuals to dried, powdered scabs harvested from the expired pustules of patients with active disease. In China it was apparently the practice to inhale this powder through the nose. In Constantinople a process called *engrafting*, or *inoculation*, was employed. The powdered scab (or liquid pus, depending on the practitioner) was rubbed into scratches made in the skin. The treated individual would develop mild symptoms approximating the first stages of the disease but would generally recover without further incident. Unfortunately, some people developed full-blown disease, and occasionally some died, as a result of this process. But those who went through it successfully were in fact quite resistant to smallpox for many years, if not for life. This is the very first instance we know of in human history involving the controlled induction of resistance to a natural disease process. Also contained within this procedure was a remarkable fact not fully appreciated until near the end of the nineteenth century—namely, that disease could be passed from one person to another by a physical agent. Probably because such a notion did not fit in with any contemporary view of disease, it was simply ignored.

One of the more colorful figures involved in bringing the practice of inoculation to the West was also one of the most brilliant women of eighteenth-century England—Lady Mary Wortley Montagu. Born into an aristocratic family in 1689, as a young dark-haired beauty she eloped with the equally aristocratic Edward Wortley Montagu, a Whig member of Parliament, in 1711. The dazzling young couple quickly became part of the court of the Hanoverian King George I. However, after barely a year in London, Lady Mary was stricken with the smallpox, and although she rapidly recovered, her beauty was ever after marred by a pocked skin and the loss of her eyelashes. In spite of this personal setback, which we know from her letters and other writings was devastating for her, she went on to become an outstanding essayist, a quite competent poet, and an absolutely delightful commentator on virtually every aspect of the society in which she lived.

Several months after her recovery from the disease, her husband, Edward, was posted to Constantinople as the English ambassador. After their arrival there in 1717, perhaps because of her own recent and traumatic brush with smallpox, Lady Mary became interested in the Turkish practice of inoculation. Impressed by what she observed, she wrote a now-famous letter to her friend Sarah Chiswell, from which the following is excerpted:

> The small-pox, so fatal, and so general amongst us, is here entirely harmless by the invention of *ingrafting*, which is the term they give it. . . . I am patriot enough to take pains to bring this useful invention into fashion in England; and I should not fail to write to some of our doctors very particularly about it, if I knew any of them

that I thought had virtue enough to destroy such a considerable branch of their revenue for the good of mankind.

As her remarks imply, Lady Mary was no great fan of the medical establishment in eighteenth-century England. But she was sufficiently convinced by what she saw in Turkey that she allowed her own six-year-old son to be treated by this method before the family left Constantinople. Her personal physician, a Scot named Charles Maitland, oversaw the procedure, carried out with the assistance of a local woman practitioner. The embassy chaplain roundly criticized the practice as "un-Christian" and would have nothing to do with it. The boy developed what appears to have been a slightly stronger than usual reaction, with high fever and numerous pustules (the latter left no pocks, however). But in all other respects the treatment was a complete success, and the young Montagu enjoyed long-lived protection from smallpox. Lady Mary did not have her younger daughter inoculated at that time, because the girl's nurse had never been exposed to smallpox; there was a general awareness in Constantinople that individuals inoculated with smallpox could pass it on to others during the active stages of the disease-like process the inoculation induced.

Several years after the Montagus returned to London, another of the periodic smallpox epidemics broke out, this time a rather severe one. Lady Mary quickly decided to have her daughter treated by the same method that had imparted smallpox immunity to her son. She summoned Maitland back from semiretirement in a village just outside London and entreated him to undertake the procedure. Maitland was somewhat reluctant to do so in "civilized" England; he agreed to do it only if several other physicians could be present as witnesses, both at the inoculation and for the follow-up stages. Lady Mary thought this a bit fussy at first—she generally disliked doctors and preferred not to have them prowling about her home—but she finally relented and the procedure took place as planned. The treatment was again successful, and both of her children enjoyed lifelong immunity from the smallpox.

The actions taken by Lady Mary, together with increasing interest in the process of inoculation on the part of the medical and scientific establishment in London at the time, led to a bizarre but interesting subsequent incident. The royal family, in particular Caroline, Princess of Wales, was gradually moving toward the possibility of having the various royal offspring inoculated. It is not entirely clear how the royals came to know about the process. Lady Mary and Princess Caroline were probably at least speaking acquaintances, and are thought to have corresponded. In addition, the princess's own personal medical advisors were also well aware of this procedure. However it came about, the royal family decided that before exposing their own children to this still uncertain practice, it would be prudent to have a firsthand demonstration of its efficacy. Accordingly, as we read in a London newspaper of June 17, 1721:

> A Representation having been made to his Majesty, by some Physicians, that the Small-Pox may be Communicated by insition or inoculating, as some express it, and that it has been practic'd safely and with Success, as might be experienc'd if some proper Objects to Practice on, were found out: 'Tis assured that two of the Condemn'd Prisoners, now in Newgate, have, upon this Occasion, offer'd themselves to undergo the Experiment, upon receiving his Majesty's most gracious Pardon. . . .

And so, on an August morning in 1721, what Arthur Silverstein has called the "Royal Experiment" began. Six prisoners, rather than the two originally envisioned, were brought up from the prison. Three men and three women, aged nineteen to thirty-six years, all of whom had been condemned to death by hanging, became the center of attention for the three physicians actually carrying out the procedure, and another score or more physicians, scientists, and other notables present as observers. The press was either present, or at least was duly informed, for the proceedings were described in some detail in local papers a few days later. In fact, the press not only reported quite regularly on the progress of the experiment, but quoted or reprinted a number of letters and articles relating to this novel Turkish import.

After the brief procedures were completed, the prisoners were held for observation for a few weeks, and then released as promised. All had recovered after developing the usual mild symptoms. But the attending physicians, doubtless with an eye to future inoculations involving royalty, wanted further evidence that they had in fact imparted immunity to their experimental subjects. One of the prisoners, a nineteen-year-old woman named Elizabeth Harrison, was induced to go out and nurse people with active smallpox, an almost certain invitation to contract the disease. She was even made to lie in bed at night with a ten-year-old boy at the very peak of his disease. She never developed the slightest symptoms of smallpox.

That may have been enough for the physicians and scientists, but Princess Caroline apparently wanted even more reassurance. She proposed using the entire orphan population of St. James Parish as a further test. In the end only a half dozen or so children were inoculated, but again with good success. Finally, in April 1723, she committed her two children, Princess Amelia (aged eleven) and Princess Caroline (aged nine), to the procedure. This too was successful, and it was naturally covered extensively in the British newspapers. There followed a wave of inoculations among many members of the apparently reassured and ever imitative upper classes in London and elsewhere throughout the country.

How could something like the "Royal Experiment" have ever happened? Why didn't anyone cry out that this was immoral, unethical, inhumane? The sense of indignation that we feel about what happened in this case may be a classic example of what has been called "presenting"—the application of present mores and standards to things that happened, in this case, almost three centuries ago. The notion that prisoners and orphans are basically property of the state, and can be coerced into medical experimentation, may seem outrageous today, but it did not raise a single eyebrow in eighteenth-century England. Some concern was expressed about condemned criminals being released back into society, but no one questioned the appropriateness of the experiments themselves. The king did take the precaution of consulting his attorney and solicitor general, who rendered the opinion that "the Lives of the persons being in the power of his Majesty, he may Grant a Pardon to them upon Such Lawful Condition as he shall think fit." It was deemed perfectly appropriate to risk the lives of condemned prisoners for "the Generall Benefit of Mankind." The rationale for the subsequent experiments on orphans was not spelled out as precisely, although the royal surgeon declared "What I thought proper to urge was, that these fresh instances might reconcile those that were yet diffident about the success of inoculation." Indeed.

True, this was not idle or specious experimentation set up for the entertainment of the aristocracy; there was plenty of evidence in hand that inoculation was both safe and efficacious, and it could readily be imagined that everyone subjected to it would benefit from it. But there was clearly contained in the Royal Experiment the notion that the lives of persons of the lower social orders of the times were less valuable than the lives of the upper classes, a notion that continued well into this century in most parts of Europe and to some extent in the United States. Medical ethics is really a very recent development. Today, years of investigation using animals would have to precede even the suggestion that any such procedure be tried on a human being, of whatever rank or station in life. A detailed description of a proposed clinical trial would have to be submitted to a Human Subjects Protection Committee and would have to contain provisions for fully informing the patient of the risks and benefits of the proposed procedure. One can only wonder what went through the minds of the six prisoners as the reddish powder was rubbed into their wounds that August morning in 1721. With the shadow of the hangman standing long on them, their thoughts must have been complicated indeed. Detailed legal and medical explanations of the "risk/benefit ratio" probably would not have figured large in their final decision to cooperate.

Although inoculation was enthusiastically embraced by the upper classes at the time, in fact it never really caught on with the general population, either in England or any other country. This failure certainly had nothing to do with the efficacy of inoculation. The Royal Society (the precursor then and equivalent now of our National Academy of Science) carried out a detailed study of inoculation in the years immediately following the Royal Experiment. Of 897 persons inoculated between 1721 and 1727, only 17 (a ratio of 1 in 53) died from what were presumed to be complications of the procedure. This certainly compared favorably with the death rate from natural smallpox—about one death in every six cases. Of a total of 218,000 deaths recorded from all causes in England during that same period, 9 percent were from smallpox. Thus from a public health point of view, inoculation made great sense.

But it didn't necessarily make common sense to the population at large, for a number of reasons. First, the procedure was not entirely safe, and a great deal was made of this within, but especially outside of, the medical community. Among scientists and physicians, it would seem that a large majority believed that inoculation was a justifiable and recommendable procedure, given the statistics just quoted, yet a substantial number of physicians forcefully opposed it. Some were concerned about the risk, and not completely convinced that the protection was genuine or long lasting. Others were concerned that the method as practiced had potential for actually spreading the disease (which it clearly did!). Still others likely opposed it for reasons as much religious as medical. And in the end there may have been a bit of class consciousness involved: Consider the following from a letter written by the physician William Wagstaffe:

> Posterity perhaps will scarcely be brought to believe, that an Experiment practiced only by a few Ignorant Women, amongst an illiterate and unthinking People, shou'd on a sudden, and upon a slender Experience, so far obtain in one of the Politest Nations in the World, as to be receiv'd into the Royal Palace.

There were, nonetheless, genuine causes for concern, which, although not obviously (now) always directly attributable to inoculation per se, caused a great deal of genuine concern at the time. One case, published by the apothecary Francis Howgrave in 1724 as a letter to Dr. James Jurin, Secretary of the Royal Society, must have been particularly discomfiting to the medical establishment, since it involved a patient handled by none other than the surgeon to Princess Caroline. The nine-year-old daughter of one "Mrs. Anne Rolt" was inoculated, and at first seemed to undergo a normal set of reactions to this procedure. But for whatever reason, she did not fully recover, and as her mother testified, "In nine weeks after the Inoculation, and after the most miserable suffering, that ever poor creature underwent, she died worn to nothing but skin and bone. She had six and thirty running sores (none of them having ever been heal'd) when she died; and they were forc'd to roll up her joints in pastboard, least the joynts should fall out of their places."

This is clearly grist for the antiestablishment press of any age, and much was made of such cases by many newspapers of the time; particularly lurid descriptions were often printed up and sold as separate tracts. There is simply no way to know whether the suffering this poor child went through was related to her inoculation or not. People who die of smallpox rarely live nine weeks. They usually have more than thirty-six sores. Was the mortality associated with this procedure (or others that ended badly) because inoculation was itself a priori dangerous? Or could there have been complicating factors? Lady Mary Montagu was quite convinced that many of the physicians carrying out inoculations did not know what they were doing, creating too deep a wound, or introducing an excessive amount of scab powder. Unquestionably, mortality varied widely from practitioner to practitioner, suggesting that details of technique were quite important to the outcome. And of course no one at the time was aware of the need for sterility in making incisions or dressing wounds. The material used for inoculation was certainly far from pure; it presumably contained samples of any pathogen (disease-causing microbe) floating around the system of the donor at the time it was collected, plus any that may have crept in while it was stored and transported. Whatever the source of the occasional problems, even its staunchest defenders admitted there was a definite risk associated with inoculation. But in general they recognized that simply being alive in eighteenth-century England represented an even more substantial risk of dying prematurely from the smallpox.

The clergy also expressed strong opposition to inoculation. Some were undoubtedly concerned about the safety of the procedure, but for others there were more fundamental issues involved. In July 1722, nearly a year before Princess Caroline finally had her daughters inoculated, the Reverend Edmund Massey delivered a sermon at St. Andrews Church in Holborn, in which he spoke out strongly and eloquently against inoculation. He sounded a theme that would be picked up, expanded, and refined by others, both in England and America, until well into the twentieth century. Rev. Massey proclaimed that the imparting of disease to an individual is done by God, at the pleasure of God, often as a punishment for sin, or to test the individual's faith. Human beings, he declared, had no business meddling around with and possibly opposing divine providence. Clergy ministering to the lower classes sounded this message with unusual force and venom, describing inoculation as heathenish and the work of the devil, consigning to eternal damnation both those who practiced it and those who submitted to it. Yet, in all fairness it must be noted that some clergymen were among the strongest supporters of inoculation, most notably Cotton Mather in colonial Boston, who (despite his previous record of opposing almost anything new and different) worked tirelessly at every level to promote widespread adoption of inoculation for the prevention of smallpox. But the clergy in America were in general even more fanatically opposed to inoculation than their English cousins, possibly because the procedure was known to have been practiced by African slaves, who brought it with them from their homelands. In the end, the practice was not adopted any more readily in the New World than in the Old.

The practice of inoculation gradually fell off by the 1730s, rising again from time to time when there would be an outbreak of smallpox. Several attempts were made to extend the benefits of inoculation to a wider segment of the population, particularly to the poor. In 1743 the governors of the Foundling Hospital decided that all children in their institution who had not previously contracted smallpox should be inoculated as a condition of being treated in their hospital. This was the nearest thing to a program of compulsory inoculation that developed. The involvement of government in promoting public health measures was an untried idea in the eighteenth century. The Smallpox and Inoculation Hospital was founded in 1746, but treatment was entirely voluntary. Separate wings were maintained for treatment of natural smallpox infections and for administration of inoculations. Such hospitals as these were supported by wealthy individuals and almost always had arrangements for treating the poor without charge. People from the upper classes usually had the procedure done in their home.

In the end, smallpox inoculation as it was originally practiced in England and America probably failed to catch on because people were simply not ready for it. We must remember that this was the first time in human history something like this had ever been tried. Doctors might practice it, but they were at a complete loss to explain to themselves or anyone else how it worked. Its origins among people considered somehow inferior tainted it in a way that even the imprimatur of the prestigious Royal Society could not overcome. And finally, as one of the more sensible clergymen of the day observed, fate and guilt mix in strange ways in humans. One could hope that one's children might be exposed to a "favorable" case of the smallpox, recover, and be protected for life from this dread disease. If it was not favorable, and the child died, one could always find solace in one's faith in God's purpose, and in the fact that death from the smallpox was simply part of life for everyone in those days. But if an otherwise healthy child died as the result of deliberate inoculation with the smallpox, under the parents' urging and direction, what then? How could one live with that? Perhaps better to leave such things alone, and take one's chances with divine providence.

Before leaving the story of smallpox inoculation, we consider one last, rather fascinating anecdote. Disease has long been a major mortality factor in war, frequently changing the course of a battle or an entire campaign. Prior to World War II, military losses from disease were always greater than losses from battle wounds. The armies of Alexander the Great were decimated on more than one occasion in the fourth century B.C. by smallpox. During the American War of Independence, smallpox was, in army camps as in the poorer sections of crowded cities, a serious problem. But it was a more serious problem for the colonials than for the British. By far the largest number of British troops came from London, where smallpox was endemic in the seventeenth and eighteenth centuries. Most young men of military age had already been exposed naturally to the smallpox, and were thus immune to it. Moreover, starting in about the 1750s, the British army began routinely inoculating recruits who had not had smallpox or previously been inoculated for it. American military recruits and volunteers, on the other hand, tended to come more from the countryside. They were much less exposed to natural smallpox, and even fewer had been inoculated. General George Washington and other military leaders were concerned throughout at least the early parts of the war about the devastation of manpower caused by outbreaks of smallpox in the army camps.

There was one campaign in particular in which smallpox played an important role, and which may have changed forever the political map of North America. In 1775, fearing a penetration of British forces from the fortress of Quebec down into New York State, the Americans sent a sizable contingent of about two thousand troops and irregulars to attack the British defending Quebec. As it turned out, the fortress was only lightly defended when the Americans arrived. The British governor gathered what troops he could for a makeshift defense and was fearful he would not be able to hold out. But shortly into their siege of Quebec, smallpox broke out among the colonials. It is interesting to speculate where the disease came from. Smallpox has a short incubation period; anyone who had been infected prior to the start of the campaign should have exhibited the disease long before the troops arrived at Quebec. At any rate, given the cramped conditions of the colonial bivouac, the tired condition of the troops, minimal nutrition, and primitive sanitary conditions, the disease swept through the colonials like a scythe. Over half the soldiers developed the disease, and the mortality rate was extremely high. No one on the British side seems to have been affected. Morale among the Americans quickly degenerated, and they withdrew to a military outpost at Lake Champlain, where generals and privates alike continued to die at a high rate in one of the most serious smallpox outbreaks of the war.

In the meantime the British rapidly reinforced Quebec with enough troops to make it a much less inviting target, and in fact the Americans never again made a serious attempt to intrude into the region. Of course we can never know for sure, but historians have speculated that had Quebec fallen to the colonialists, and been reinforced and maintained as a strong American outpost, much of the eastern part of Canada would today be part of the United States. And, one supposes, the question of francophonic separatism would be an American problem!

Vaccination: The End of a Plague

The next advance in active immunization against smallpox came just a few years later, at the turn of the nineteenth century. Toward the end of the eighteenth century, people began to notice that individuals exposed to the *cowpox* seemed to be highly resistant to infection with smallpox. We now know that cowpox is caused by a virus—called the *vaccinia* virus—that is closely related to the smallpox virus. In fact, the two viruses are more than 95 percent identical, with slight differences in probably no more than a dozen genes out of several hundred. In milk cows, vaccinia causes an eruption of blisters on the udders that clear up quickly and induce no serious illness. Farm workers who come into contact with cows during the peak of infection may develop a reaction similar to the mildest forms of smallpox: a bit of fever and achiness, with the eruption of a few blisters on the hands and lower arms that occasionally develop into depigmented pocks. Once exposed to the cowpox, however, neither cows nor humans develop any noticeable symptoms upon subsequent contact with infected cows.

The relation between cowpox and smallpox may have been "known" in at least a folkloric fashion for many years prior to the eighteenth century. As with folklore generally, the origins of such a connection are hard to trace. Among urban intellectuals of the day who actually thought and wrote about such things in the early part of that century, there is no indication that they were aware of a connection between the two diseases. It is possible that they may have heard rumors of such a connection but dismissed these as the fevered imaginings of country folk, whom Londoners did not hold in particularly high esteem. But beginning in about 1760, it is clear from various records that physicians and scientists were beginning to discuss such a connection among themselves. What had likely been noticed with the advance of inoculation into the countryside was that dairy workers who had had cowpox did not develop the usual set of smallpox-like symptoms when inoculated by the standard procedure with smallpox material. Of course, once this connection was convincingly demonstrated, many people claimed to have known about it all along. Some even claimed to have purposefully exposed themselves or their children to cowpox as a precaution against the smallpox. It is hard to say this was absolutely not the case, but with one or two possible exceptions there is little to support it.

The first person to investigate seriously the relationship between smallpox and cowpox was Edward Jenner. Jenner was born in 1749, at a time when smallpox inoculation was becoming increasingly safe and effective but, paradoxically, practiced less and less frequently. Jenner is often described as a physician, but in fact he was what was then called a "surgeon-apothecary," which at least socially was quite a step below a true physician, or one who practiced the "Physick." The title of physician could be claimed only by someone with a university degree in Physick—an M.D. Surgeons were admitted to medical practice after a period of apprenticeship to a practicing physician or surgeon, without certification. Men (for it was only men at the time who were admitted to medical practice) could achieve professional eminence and distinction with either title, but the social distinction was clear. Jenner had, however, been made a fellow of the Royal Society in 1790, based not on his medical accomplishments but in recognition of his studies of the nesting habits of the cuckoo bird.

It is not obvious that Jenner's interest in the cowpox was related, at least initially, to its alleged ability to impart immunity to smallpox. He is known to have been interested in the possibility that certain diseases found their way into the human race from the animals that early human societies had domesticated. This idea shows up in his writings before and after his discoveries relating cowpox to smallpox through immunization. Given the commonly held belief that diseases are one of God's ways of punishing sinners, Jenner's thoughts on this subject were not likely to have been popular at the time. But in fact modern-day anthropologists and pathologists would probably agree with many of his ideas, in broad outline if not in the specifics. Jenner ultimately came to believe that cows had picked up their pox disease from horses and that humans very likely acquired theirs from cows at some time in the distant past. Jenner apparently reasoned that if human smallpox was related to cowpox, then immunization with the latter might provide resistance to the former.

Like the royal family before him, Jenner felt perfectly free to experiment on live, healthy human beings to test out his ideas. In mid-May 1796, he collected pus from an active cowpox sore on the hand of a milkmaid, Sarah Nelmes, who worked on a nearby farm. Using the by-then standard technique for inoculating with smallpox material, Jenner made superficial scratches in the arm of an eight-year-old boy, James Phipps, and rubbed in some of the liquid pus collected from Miss Nelmes. The boy developed a slight fever a week or so later, but apparently he did not develop any obvious blisters or lesions on his body. Jenner recorded in his notes that it looked pretty much like a mild reaction to a successful smallpox inoculation. Then, just under seven weeks later, he inoculated young Phipps with matter taken directly from an active smallpox pustule. The material was rubbed into several scratches and punctures on both arms. One can only imagine the intensity with which little Jimmy Phipps must have been observed over the next two weeks. Smallpox inoculations were fairly safe by the end of the 1700s, but never completely so. If the boy had died in the course of Jenner's experiment, there surely would have been the devil to pay. The anti-inoculators would have had a field day, and rightly so. Fortunately for Jenner (and for all of us, in the end), young Phipps developed no reaction whatever to his inoculations.

To Jenner's mind, the case was proved. He wrote up a description of his experiment, together with a number of observations he had heard about or witnessed concerning the likely protective effect of the cowpox vis-à-vis smallpox, and submitted it to the Royal Society. To his great dismay the Society declined to publish his paper, so a few years later he published it as a pamphlet at his own expense. In its revised form it contained reports on a few more cases that had come to his attention, and it included a second experiment in which he had inoculated a five-year-old boy, William Summers, with pustular material taken directly from an infected cow udder. This too rendered the recipient resistant to a subsequent infection with smallpox material.

The reaction to Jenner's pamphlet was slow at first. Jenner was not part of the medical elite of his time, and London physicians tended to dismiss him and his ideas. Cartoons appeared in the popular press showing children with cow horns growing from their heads as a result of Jenner's procedure. But within a year or so several highly respected physicians began using the "Jennerian technique" (which came to be called *vaccination*, after the Latin *vacca*, cow). Results compared favorably to previously used methods for inoculating against smallpox, now referred to as *va-*

riolation. By the turn of the century, the advantages of vaccination over variolation were becoming quite clear, and Jenner's fame grew. Study after study confirmed that excellent protection against smallpox could be obtained by inoculation with cowpox, with minimal morbidity and almost no mortality. Today we know that is because the virus that causes cowpox is very similar to the one that causes smallpox; protection against one confers protection against the other. At the time, the fact that cowpox inoculation was embraced by members of the socially approved medical establishment was probably the dominant factor in its acceptance.

In 1802, Parliament awarded Jenner a prize of £10,000, a substantial sum of money. The award was supplemented in 1806 with another £20,000. He also received numerous honors and degrees from institutions and governments around the world. To help promote the practice of vaccination, Jenner's friends petitioned the king to allow formation of a Royal Jennerian Society; this was established in 1803 with the queen's personal patronage. He was admired by Napoleon, who had a medal struck in Jenner's honor and released several British prisoners in response to pleas from Jenner. But Jenner never was given a knighthood by his own country, nor was he "legitimized" by being made a Fellow of the Royal College of Physicians.

The Jennerian technique for immunizing against the smallpox was carried forward with only minor modification, and continued in use until the final eradication of smallpox in the middle of the twentieth century. Especially effective strains of cowpox were nurtured in various places around the world, and the precious pus was collected and stored for future use. There were continued improvements in collection and purification procedures to make the vaccines safer, but the technique until the very end remained essentially Jenner's. Whatever his reasons for getting into the inoculation game in the first place, and whatever errors in judgment he may have made along the way, unquestionably Jenner has been vindicated by history in a way granted to few individuals.

Why was Jenner's technique accepted, whereas variolation had met so much resistance and hostility? In fact, the very same forces that opposed variolation opposed vaccination as well. Arguments from the medical and scientific communities (some of whom by that time had a considerable professional and personal investment in variolation) dropped away fairly early in the face of overwhelming statistical evidence in favor of vaccination. But such considerations had little impact on those who argued against vaccination on purely religious, ethical, or simply antiestablishment grounds.

In the first fifty years after the discovery and confirmation of the Jennerian technique, the British government began not only to encourage vaccination but also to make it compulsory through the passage of various laws and acts. This led to enormous controversy, touching on issues of civil liberties and the proper role of government in health matters that are still with us today; think of the controversies related to fluoridation of water, or cancer "cures" such as laetrile or Krebiozen, or experimental drugs for AIDS. Initial resistance had a clear basis in the sketchy reproducibility of vaccination. Cowpox pus samples varied widely in their efficacy, since no one really understood what factors in pus collection were critical. This was worked out entirely by trial and error over the years. Methods for handling patients both before and after the procedure varied widely among different vaccination centers, and between wealthy and poor patient populations. A decade or two after the start of vaccination, it became apparent that many people immunized by vaccination were still susceptible to smallpox. Part of this was certainly due to inadequate technique in the early years, or faulty vaccines. But the realization gradually crept in that even the best vaccinations might not confer lifelong protection against smallpox. Thus was born the concept of the "booster shot," or secondary immunization, which today is quite common and which makes perfectly good sense in terms of what we understand about immunological memory.

Unfortunately, at the time it seemed that every such incident was seized upon by opponents of vaccination as a further example of the evil inherent in the procedure itself. Religious leaders, particularly among the working classes, still held that smallpox was part of God's armamentarium for punishing the wicked, and that both vaccination and variolation interfered with divine providence. Others felt that smallpox and other diseases could be eradicated simply by improved public hygiene. Undoubtedly, the great moves forward in public sanitation in the nineteenth century were as much responsible for improvements in general health and life span as anything else that happened, including vaccination. But there was a viciousness in the opposition to vaccination that, while difficult to comprehend, is important to recognize, for we see it over and over again in human affairs. Perhaps it is one of the few ways that the unempowered of any age have of making their weight felt. Consider the following statements by the American J. M. Peebles, trained as both a physician *and* a scientist, whose child apparently was refused admission to an elementary school for lack of a required vaccination certificate. Peebles subsequently (1900) wrote a book entitled *Vaccination: A Curse and a Menace to Personal Liberty, with Statistics Showing its Dangers and Criminality*, which became a major sourcebook for the antivaccination movement.

> The vaccination practice . . . has not only become the chief menace and gravest danger to the health of the rising generation, but likewise the crowning outrage upon the personal liberty of the American system. . . . Compulsory vaccination, poisoning the crimson currents of the human system with brute-extracted lymph under the strange infatuation that it would prevent small-pox, was one of the darkest blots that disfigured the last century.

Fortunately, the forces of reason prevailed, and the practice of vaccination, when extended to pathogens other than the smallpox virus, would reduce enormously the morbidity and mortality to humans caused by infectious diseases.

So whatever happened to the smallpox virus itself? The last natural case of smallpox referred to earlier occurred in Somalia in April 1977. A young hospital cook named Ali Maalin had helped transport a young girl with smallpox and ended up contracting the disease himself. The girl may have been the last human on earth to die of a natural smallpox infection; Ali Maalin recovered, and in fact is still alive today. The World Health Organization (WHO) declared smallpox officially eradicated from the face of the earth on October 26, 1979.

But the Somali child would not be the final victim of the smallpox virus. Just one year later, in July 1978, a medical photographer named Janet Parker working in a Birmingham (England) hospital became infected by a strain of smallpox vaccine that seems to have passed through an air duct from a laboratory on the floor below. She died two months later. Everyone who had come in contact with her was vaccinated and followed closely, but she was the only person to become infected. The director of the laboratory from which the virus originated was so despondent over this incident that he eventually committed suicide. This case, and several other grim accounts of near tragedy with lab stocks of smallpox virus, have led the WHO to appoint a committee to study whether the last remaining stocks of smallpox virus should be destroyed. The cowpox virus would of course be exempted from any such death sentence. Thus the last few vials of the virus that has killed uncountable millions of human beings sit in their liquid nitrogen cells in Atlanta, Gerogia (the Centers for Disease Control), and in Moscow (the Institute for Virus Preparation), awaiting the final word. Will it be life in the deep freeze or death in the autoclave? We should know soon!

Beyond Vaccination: Pasteur, Koch, and the Germ Theory of Disease

Despite opposition from many quarters, vaccination moved forward in the first half of the twentieth century, and the death rate from smallpox began to decrease slowly but inexorably in every society in which vaccination was officially adopted. But what about other diseases? What did we learn from smallpox that could help conquer other human maladies? Frankly, not much. Inoculations for smallpox consisted of gathering material from sores and using it to immunize an otherwise healthy person. That's fine for diseases that produce sores, but what do we do for diseases that do not result in frank skin lesions? Very few diseases do. Again, we run into the fact that no one really knew the actual basis of infectious diseases. No one knew about the existence of *germs*—microscopic organisms that cause disease, and that can be passed from one person to another. This understanding was absolutely essential for the development of additional vaccines. (The term *vaccine*, although by its very name referring specifically to the use of cowpox material, was subsequently adopted for planned immunization with any disease-related material.)

Further progress in immunotherapy for infectious diseases was thus dependent to a considerable extent on progress in microbiology—the study of microorganisms too small to be seen with the naked eye. A subset of these, called *pathogenic microorganisms*, are the living agents responsible for infectious disease. They include not only viruses like those responsible for smallpox, but also bacteria, funguses, and certain parasites. The preparation of vaccines for prevention of the diseases caused by almost all of these pathogens had to await the identification and isolation of the pathogens themselves. Viruses, as pathogens distinct from bacteria, would not even be defined for a hundred years after Jenner learned by empirical means to thwart them with vaccination, and it would be fifty years beyond that before viruses in general could be prepared in forms useful for vaccination against other viral diseases.

But before this work could even begin, a breakthrough in thinking was required—one of those intellectual leaps that change the world forever. Human beings had to reach the understanding that infectious disease, whether viewed as a punishment from God or simply a rotten throw of the dice, is caused by microbes. Whether or not it is divinely inspired, disease has a rational basis. People had to realize that what was being passed from person to person, or from cow to person in smallpox vaccination, was a living thing, a thing that could be isolated, identified, studied, and, through the knowledge gained, ultimately controlled by humans. We needed a germ theory of disease.

The possibility that disease is caused by discrete, invisible entities that can be passed from one individual to another had been put forward at various times throughout recorded history but never really caught on. The famous Veronese physician Girolamo Fracastoro spelled out such a theory in great detail in 1546, suggesting that different diseases are caused by different rapidly multiplying "minute bodies" that can be passed from person to person by physical contact (including touching contaminated clothing), or through the air. Anton van Leeuwenhoek described what were obviously living organisms in his crude microscopes only a hundred years later. Thus by the mid-seventeenth century, there was a well-thought-out theory *and* physical evidence that could have supported a germ theory of disease. Why didn't such a theory occur to anyone? What took so long?

There are two major reasons why scientific theories catch on. Either they are intuitively obvious, and everyone wants to jump on the bandwagon (in which case the real fight is to slow everyone down and prevent exaggeration of the evidence); or the experimental evidence is so overwhelming that in spite of being counterintuitive, everyone finally, if reluctantly, falls into line. A germ theory of disease, to the average person throughout human history, would not at all have been intuitive. The idea that invisible living things could so fundamentally discombobulate a human being as to cause grave illness, even death, simply did not compute. Clearly this was a situation that would require irrefutable experimental evidence.

The evidence, perhaps a little soft at first, began to accumulate only in the nineteenth century. Some people, whom today we would call epidemiologists, began to study seriously how diseases spread during outbreaks, or epidemics. A close analysis of how diseases like cholera and typhoid fever spread around in crowded urban populations suggested that some sort of physically discrete, Fracastorian entity that could be directly transmitted from person to person must be involved. Without having the slightest idea what such an entity might be, early researchers proposed public health strategies for limiting the spread of diseases that were clearly based on the involvement of such agents. And in fact, such strategies seemed to work. Studies like these did not obviously lead to a *germ* theory of disease, but they did soften up the ground, so to speak, for further thinking in that direction.

Such developments were not long in coming. The major breakthroughs in identifying the agents involved as microbial life-forms came with the work of Louis Pasteur in France in the middle of the nineteenth century, and a little later with the contributions of Robert Koch in Germany. Pasteur was very likely aware of the thinking of the epidemiologists, and he was certainly aware of the existence of microbial life-forms. But the work that would lead him to the formulation of a germ theory of disease was not aimed at the study of disease per se, but rather at the process of *fermentation*.

Fermentation was of great interest in the nineteenth century to both scientists and industrialists. It was recognized that fermentation was important in the production of wine and beer, and that a basically similar process was at work in putrefaction, that is, the decay of living matter into simpler compounds. Both the souring of milk and the rotting of meat and vegetables were regarded as fermentative processes. Pasteur showed that fermentation is actually caused by living microorganisms such as microbes. He isolated microorganisms from "ferments," purified them, introduced them into fresh, unspoiled material, and caused all sorts of fermentation. In another type of experiment, he showed that fermentation could be slowed or completely halted by heat (a process we now call *pasteurization*). Once halted by heat, fermentation could be reinstated by adding back a fresh source of live microbes. Here was definitive proof that microscopic organisms were capable of causing profound changes in biological materials.

The first indication that such microbes might also be involved in disease came with Pasteur's work on the silkworm blight that devastated the silk industry in the Cévennes region of France from about 1850 on. Pasteur showed convincingly that the disease that was destroying silkworms was caused by a microbe. The presence of the microbe in silkworms was absolutely diagnostic for the disease. Pasteur could predict which silkworms would get sick before any signs of disease appeared, just by the presence of the microbe, and the disease could be imparted at will to healthy silkworms by injection with the microbe. Such a convincing demonstration of the production of a specific disease by a specific microorganism might, from a twentieth-century perspective, have been expected to cause an immediate reaction in the scientific and medical community. In fact, with the exception of some very grateful silkworm growers and several relieved government finance and trade ministers, hardly anyone noticed. Microbes were still considered essentially a biological curiosity, and the notion that lowly organisms like a silkworm could tell us anything about human disease had not yet entered the minds of even the most enlightened scientists. Today we know full well that the physiological processes of life in the lowest living organisms are remarkably similar to those in humans. But that was not at all obvious a century or more ago.

Pasteur raised the ante a significant notch with his subsequent studies on anthrax. Anthrax is a disease that, if it rages unchecked in domestic animals like cattle and sheep, can cause enormous damage. Death is rapid but agonizing in infected animals. The corpses of stricken animals bloat and then decompose very rapidly. The internal organs, especially the spleen, show extensive putrefactive decomposition at the time of death. As early as 1838, high levels of rod-shaped microbes were seen in the blood of animals dying from anthrax when examined under the microscope. But this was noted simply as a biological curiosity. The same thing was noted again by a French parasitologist named Casimir-Joseph Davaine in 1850. Eleven years later, Davaine read an article by Pasteur describing the presence of similar rod-like microbes in fermenting liquids. Davaine was struck by the apparent similarity of the microbes in the two seemingly distinct phenomena, and wondered whether the same, or at least a similar, living microorganism might be involved in both fermentation and disease. Apparently the same idea must have occurred to

Robert Koch in Germany. Both Pasteur and Koch went on to show that specific microbes could be isolated from anthrax-infected animals, grown in glass vessels (in vitro), and reinjected into healthy animals—imparting to them a fatal case of anthrax.

Surely at this point the germ theory of disease must have been proven. In hindsight, yes—but it took another ten years or so to really convince everyone. Many eminent medical authorities had based their entire careers on other theories of disease; they were not about to relinquish their beliefs. There were in fact some reasonable reservations expressed. For example, it could be shown that some of the microbes claimed to cause disease could be found in animals that were perfectly healthy. If these microbes are the sole explanation of disease, why were these animals healthy? We know now that small numbers of many potentially harmful microbes are held in check by the immune system; if the immune system is disabled, either by immunosuppressive drugs or by natural diseases such as AIDS, these crypto-pathogens suddenly come flying out of the woodwork and wreak enormous havoc. But in the late nineteenth century, this was a reasonable reservation about the new disease theory. And there was still a reluctance to accept findings in animals as relevant to human beings. A number of sarcastic comments were made about credible scientists hanging around with veterinarians. But veterinarians, then as now, were in fact well-trained scientists in their own right, and possible social distinctions of the time aside, their observations and contributions could not be just dismissed out of hand.

And finally, Robert Koch would isolate the first microbe to cause a disease—tuberculosis—in humans just a few years later. This would open one of the most exciting eras in scientific research of any age: the relentless pursuit and systematic control of microbes causing infectious diseases in humans. Overnight, people at all levels of society had to alter radically their views of disease, its origins, and its relationship to human life. It was a jolt of major proportions, easily of the same magnitude as the demonstration by Copernicus that the earth is not at the center of the universe. As Pasteur himself said: "If it is terrifying to think that life may be at the mercy of the muliplication of those infinitesimally small creatures, it is also consoling to hope that Science will not always remain powerless before such enemies. . . . All is dark, obscure and open to dispute when the cause of a phenomenon is not known; all is light when it is grasped."

The rapid acceptance of the germ theory of disease was due in a very large part to the work of just two men: Pasteur and Koch. These two, more than any others of their age, opened the golden frontiers of microbiology and immunology that led in a few short decades to the control, if not the complete eradication, of the major infectious diseases that had been the scourge of humanity from the beginnings of history. Unfortunately, rather than working together, these two men were caught up in the orgies of nationalism that preceded and followed the Franco-Prussian War. Although the war lasted less than a year (July 1870 to May 1871), it was attended by feelings of bitterness that extended to every level of German and French society. Science and medicine were not exceptions, and men who were otherwise paragons of politeness and propriety became, at scientific meetings, little more than street brawlers in the cause of their respective national honors.

Pasteur and Koch were very different personalities, each in a way reflecting his nation's stereotype: Pasteur—warm, effusive, personal, emotional; Koch—cerebral, aloof, precise, never one to suffer fools gladly. In 1868 the University of Bonn, in recognition of his studies on fermentation, had conferred on Pasteur an honorary degree of Doctor of Medicine, attended by sincere praises for his contributions to medical science. Less than three years later, after the German occupation of Paris, Pasteur returned his degree with an angry letter, saying that he found the presence of his name and the name of the German head of state on the same piece of parchment "odious"; he asked the university to "efface my name from the archives of your faculty, and to take back that diploma, as a sign of the indignation inspired in a French scientist by the barbarity and hypocrisy of him who, in order to satisfy his

criminal pride, persists in the massacre of two great nations."
Pasteur received a prompt reply from the president of the faculty
at Bonn, who declared himself obliged "to answer the insult
which you have dared to offer to the German nation . . . by
sending you the expression of its entire contempt." In the ensuing
decade, both Pasteur and Koch continued to make spectacular
contributions to the origin and containment of disease, but with
an underlying mutual enmity that barely allowed them to be civil
to one another in public. Koch found Pasteur pompous and technically sloppy; Pasteur found Koch arrogant and focused to the
point of narrow-mindedness. Their disciples were perfect mirrors
for their masters' vanities, and many of the publications and
international scientific meetings following the war were used as
forums for asserting national superiorities as much as for the
dissemination of scientific information.

Yet, human nature being what it is, who can say that the
personal and national antagonisms driving the competition between these two men and their followers did not but advance
more rapidly the understanding and containment of infectious
diseases that had laid waste to human beings for untold millennia. The contributions of the schools they founded in their respective countries far outweighed those from countries at peace
during that same period. The sometimes savage personal or patri-
otic satisfactions that came from identifying the next microbe or
the disease it caused may seem tasteless and petty to us today, but
that does not stop us from enjoying the protection this knowledge
affords us. Is there a lesson in all this? Undoubtedly. But are we
wise enough to understand it?

TWO

The Anatomy of
an Immune Response

A Gift of Life

On Christmas Day in 1891, a desperately sick baby girl was brought to the Bergmann Klinik in Berlin, Germany. Like many infants in Europe and America at the end of the nineteenth century, she had contracted diphtheria. The outlook for anyone coming down with this disease was not good, but it was particularly poor for small children. Diphtheria was in fact often referred to as the "strangling angel of children." It was not at all unusual for half of infected youngsters to die, whether they made it to a hospital or not. But this infant girl arrived at the Klinik at a propitious time; she was about to receive a new form of treatment that would not only save her life but also revolutionize the way in which infectious diseases afflicting human beings were perceived and treated.

Berlin at this time was at one pole of the intense rivalry that still existed between France and Germany two decades after the end of the Franco-Prussian War. This rivalry did not by any means put scientific endeavors out of harm's way, but rather used them as a weapon in the unrelenting struggle for national prestige. Research groups in both countries (as well as around the world), once the germ theory of disease was fully understood and accepted, had made rapid progress in isolating and identifying a wide range of microbes responsible for human diseases. The majority of these turned out to be a subset of microbes called *bacteria*. Bacteria are tiny and can be seen only with a microscope. About a quarter million bacteria would fit in the dot over an "i" on this page. The major question facing the new field of microbiology was how such a tiny organism could be the cause of such devastation to human beings.

Researchers in Paris and Berlin continued to make rapid progress in the 1880s in developing an understanding of how bacteria cause disease. They found that if they grew bacteria in a broth, removed the bacteria completely from the broth, and injected the broth alone into an animal, they could in many cases reproduce the disease caused by the intact bacteria. This led to the conclusion that some bacteria, such as those that cause tetanus, are able to produce and shed a chemical substance that actually causes the disease. There substances became known as bacterial *exotoxins*. And then, in 1890, two young scientists in Robert Koch's laboratory in Berlin (Emil von Behring and Shibasaburo Kitasato) published a paper that shook the scientific world. They had injected a rabbit with a dose of tetanus that was small enough that the rabbit's immune system would overcome it. They then prepared some serum from the recovered rabbit. (Serum is the straw-colored liquid that remains when blood is allowed to clot and the clotted material is removed.) When they tested this "immune serum" they found that it contained substances that could completely neutralize highly purified samples of disease-causing tetanus exotoxin. They called this new substance "antitoxin." Their most exciting finding was that when immune serum was transferred into an animal that had never been infected with tetanus, that animal was protected against a subsequent injection of a lethal dose of tetanus toxin. *Passive immunization*, as this technique came to be called, provided firm proof that the immune response to tetanus is mediated by blood-borne substances. Finally, the two researchers showed that even animals that were already infected with tetanus, and well on their way to dying, could be rescued by timely administration of the immune serum.

The implications of this reasearch for treating human disease were immediately obvious to everyone working in the area of human health. One of Koch's students had previously worked with diphtheria exotoxin; he quickly showed that the therapeutic benefits obtained using tetanus antitoxin to treat tetanus could also be obtained using diphtheria antitoxin to treat diphtheria. After a number of trials with various animals as donors of antitoxin serum, and as mock patients for antitoxin serum therapy, they decided (under the strong urging of their superiors) to make the leap from animals into humans. Their Christmas gift to the world for 1891, from the Bergmann Klinik in Berlin, was the life of a young girl dying from diphtheria.

Even at this distance in time, the excitement and wonder this accomplishment generated still brings a silent rush to any scientist who has wondered whether the long hours he or she has spent in the lab will ever matter, will ever be noticed or understood by more than a select handful of peers. The techniques of vaccination had been steadily improved upon in the decades since its introduction. It was also, slowly but surely, beginning to be extended to diseases other than smallpox. But the question nagging even the most ardent supporters of vaccination remained: What is the change wrought inside the body as a result of vaccination? What is the nature of immune protection? Now they knew; it was a substance shed into the bloodstream. Its isolation and identification would only be a matter of time.

Without waiting for the magic factor to be identified, the German government immediately supported large-scale efforts to produce antisera against a wide range of bacteria and bacterial toxins causing human disease. Clinical trials for testing antisera directly on human beings were organized throughout the country. Similar programs were soon instituted in other European nations and in the United States. Death rates from diseases like diphtheria dropped almost overnight, although, it should be noted, not to zero. Most institutes reported decreases in mortality on the order of 50 percent. Many problems had to be worked out before the injection of animal antibodies into humans would be even relatively safe. But in the context of deaths caused by infectious diseases at the end of the nineteenth century, this was a major step forward.

One of the young scientists who participated in the discovery of antitoxin therapy, Emil von Behring, would eventually receive the first Nobel Prize in Medicine awarded, in 1901. The astounding success of this young German army doctor-cum-scientist may have signaled the end of the bitter national rivalries that had characterized the preceding quarter century. In 1895 he was awarded the prestigious Prize of the Académie de Médicine de France. When he died in 1917, von Behring was eulogized in a prominent British medical journal, which did not even mention the fact that he was a citizen of a country at war with England.

In a sense, the demonstration that antitoxins (which are now referred to by the more generic name *antibodies*) could rid the body of disease was the culmination, the ultimate payoff, of hundreds of folk observations and laboratory experiments, all the way from the women who practiced variolation in the Middle East, through Jenner, and on into the titanic battles waged between Pasteur and Koch. But if it was an ending point, it was also a beginning—the beginning of the field of immunology. The possibility that there might be a special system in the body to protect us against disease, and that this system could be manipulated to protect us in the absence of the disease itself, attracted thousands upon thousands of scientists and doctors to this new discipline.

What would be learned in the 100 years that followed is that there are in fact two major arms to the immune system: the antibodies described by von Behring and Kitasato, and a second and equally powerful defense called *T cells*. These systems were obviously designed to stand between us and the uncounted hordes of microscopic pathogens that would like to subvert our bodies to their own ends. Only later would we come to realize that the immune system—these very same antibodies and T cells—are truly a double-edged sword, with the potential to harm as well as to help. That was an uncomfortable and confusing notion, not readily understood in the heady early days of immunology. It was, understandably, easier and more satisfying to concentrate on the positive and nurturing aspects of the immune system, which we will now explore briefly.

The Antibody Response

Any foreign substance of a biological nature (which immunologists refer to as *antigen*) when injected into humans elicits the production of special proteins called *antibodies*. The modifier "biological" is not placed here idly: The immune system does not waste time and energy making antibodies to nonbiological materials. Our bodies have learned during millions of years of evolution that inanimate substances are rarely harmful. The real threat comes from other living things, things like bacteria, viruses, fungi, and parasites that want to live and reproduce inside us. Those are the antigens that bring the immune system to a full state of alert. Antibodies are expensive to make in terms of biological energy; they cannot be expended against meaningless threats.

Antibodies are produced by a special white blood cell called a *B lymphocyte*, or simply a *B cell*. Antibodies appear in the bloodstream about three days after the first encounter with a given antigen. Once made, antibodies circulate throughout the body in search of the antigen that triggered their formation in the first place. When they find the antigen, they bind tightly to it, which triggers a series of events leading to removal of the antigen from the body.

Although antibodies are energetically expensive to make, the immune system does not skimp on the number of different antibodies it is prepared to produce against something foreign, as long as that something foreign is biological in nature. The immune system cannot afford to be stingy. The number of different forms of life that can live within us, causing disease and even death, is enormous. Moreover, these life-forms are able to mutate and change themselves at rates far in excess of the rate at which we can make corresponding changes. Thus our immune systems are able to produce a huge number—certainly more than a hundred million—different types of antibody to deal with the "antigenic universe." This feature of *diversity* was one of the earliest recognized hallmarks of the immune system.*

Antibodies have the special property of binding specifically and tightly to *only* the antigen that induced their formation. This property of *specificity* is another important feature of the human immune system. An antibody against the smallpox virus, for example, does not react with and lead to the elimination of diphtheria toxin, or vice versa. But most important, antibodies produced by B cells in response to foreign antigen also do not react with *self*, that is, with our own cells and tissues. Thus a third important characteristic of immune responses is *self-tolerance*. It is perhaps one of the most difficult, yet absolutely crucial, challenges the immune system had to meet as it evolved. When the immune system fails to make this distinction properly, and begins to produce antibodies that react with the body's own tissues, the result is autoimmune disease.

Finally, the feature of the antibody response we are all perhaps most familiar with is something called *memory*. It is what we mean when we say we are "immune" to something. The first time an antigen enters the body, the response is a bit slow, and not very strong. It takes a while for the immune system to gear up against something completely new, something it has not seen before. But once the immune system has learned to recognize and eliminate a foreign antigen, it rarely forgets. The next time the same antigen appears inside the body, the response is swift and overwhelming. In fact, the more often the immune system recognizes a given antigen, the faster and stronger the immune response is.

These are the four cardinal features of the antibody response:

---

*The unraveling of the way in which the immune system responds to the enormous range of different antigens in the antigenic universe is one of the most exciting chapters in the intellectual development of immunology. This story is told in more detail in the appendix.

diversity, specificity, self-tolerance, and memory. The same properties are characteristic of the second branch of the immune system: T cells.

T Cells: The Second Arm of the Immune Response

The suspicion that antibodies might not be the entire explanation of how we respond immunologically to foreign antigen first arose in the 1940s, but lay around largely unconfirmed until the early 1950s. The story of the discovery of a second arm of the immune response is one I always delight in recounting for students because it illustrates so beautifully how "real science" is often done.

Bruce Glick, a graduate student at the University of Ohio in the 1950s, had become interested in a small sac at the tail end of the digestive tract in birds called the *bursa of Fabricius*. In anatomy, if the function of a structure is unknown, it is usually just given the name of its discoverer. This particular structure was obviously a real puzzle—first described by Hieronymous Fabricius in the sixteenth century, it had never been renamed. Glick tried the time-honored approach of simply removing the bursa from chickens of various ages, including newly hatched chicks, and waiting to see what would happen. After a variety of experiments of this type, he could find no obvious differences in chickens with or without a bursa. He finally gave up and returned all his chickens to the general stock.

Enter another graduate student, Tony Chang, a teaching assistant in need of a few chickens to demonstrate the production of antibodies. To save money, Chang selected Bruce Glick's chickens for his demonstration, including the ones that had been bursectomized at a very young age. (The timing of the bursectomy, as it turned out, was critical.) To Chang's embarrassment in front of his class, the animals failed to produce antibodies. Now, at this point, many graduate students would have just shrugged and eaten the chickens. (They are used to experimental failures, especially in their early years, and they are almost always hungry.)

But these two young men put their heads together, and saw beyond the possibilities of a free meal. Together with a colleague, they carried out additional experiments that showed for the first time the important role played by the bursa in the development of the ability to produce antibodies. This conclusion, which would have been very easy to overlook, is a reminder to all students that Pasteur's maxim that "chance favors the prepared mind" still has a good deal of validity.

Together they wrote up what was destined to be a landmark paper in immunology, but the world wasn't quite ready for it yet. It was submitted to the prestigious journal *Science*, whose editorial staff rejected it as "uninteresting." It was finally accepted in the journal *Poultry Science* where, as may be imagined, it languished for some years before it suddenly became the most quoted paper ever published in that journal. Another good lesson for students: Don't let rejection by the establishment force you to give up. You may very well get the last laugh.

What made theirs a benchmark contribution to immunology was the subsequent finding by Glick and by others that while bursectomized chickens could not make antibodies, they had a perfectly normal ability to overcome viral infections and to reject skin grafts. This was a stunning finding. Both these reactions were known by the 1950s to be immunological in nature. But "immunological in nature" had always meant antibodies. By identifying a specific organ that controlled antibody production, these researchers could for the first time disable this function in an animal that was dependent on an immune system. And that in turn allowed the first look at what was left over once antibodies were turned off. The results showed clearly that antibodies are only one way the immune system has of dealing with foreign antigens. So while Bruce Glick and his colleagues are remembered chiefly as the researchers who first defined the B-cell system underlying antibody production, the major importance of their work is that it forced others to begin searching for alternative immune mechanisms. And that led to T cells.

The discovery of T cells, when it came in the 1960s, would

open a new era in immunology. There would turn out to be two major types of T cells. One type, called a *helper T cell*, is required to help B cells make antibodies to most antigens, especially protein antigens. It does this by releasing small chemical messages called *lymphokines* that are needed by B cells to mature into the antibody-producing plasma cells. Helper T cells display a molecule called CD4 on their surfaces, and are thus also called CD4 T cells. The second major type of T cell turned out to be something called a "killer T cell." The job of these T cells, which are marked by a surface protein called CD8, is to seek out and destroy cells in the body infected by viruses or bacteria, as well as cancer cells. The CD8 T cells and antibody together scour every nook and cranny of the body in search of pathogens. Antibodies detect and trigger the destruction of pathogens floating around in the bloodstream and in other body fluids; CD8 T cells detect and destroy pathogens hiding inside cells by destroying the cell they have invaded.

In trying to figure out how CD8 T cells kill pathogenically altered cells, or cancer cells, immunologists discovered a fascinating fact. For many years it was assumed that CD8 T cells must carry some sort of "weapon" they could use to kill an aberrant cell. Identification of this weapon became one of the holy grails of immunology. When years of investigation failed to turn up any such weapon in CD8 T cells, it slowly dawned on everyone that it might not be murder after all—it might be suicide. Induced suicide—"assisted suicide," if you will—but suicide nonetheless. Immunologists now believe that all cells in the body may be preprogrammed to die on command. One of the situations in which it may be advantageous for cells to die, in terms of the overall well-being of the host organism, is when such cells harbor a pathogen or have become cancerous. It has been proposed that what CD8 T cells do when confronted with such a cell is not bring out some special sort of weapon to kill it, but rather simply to punch in a "security code" that triggers a small self-destruct device implanted in each cell. This would get around the possibility of some pathogens finding ways to resist a CD8 T cell "lethal weapon"; the CD8 cell does not go after the pathogen at all; it just tells the cell to die. This mechanism of *programmed cell death* turns out to be quite widely used in the immune system, and in other systems of the body as well.

As it turns out, CD4 T helper cells, in addition to helping B cells produce antibodies, also help CD8 killer cells to mature. We will see later that the CD4 T cell is the target for the AIDS virus. That is why AIDS is such a deadly disease: Not only does the AIDS virus cripple CD4 cells, but through them, it also abolishes both the antibody and killer T-cell arms of the immune response.

The Lymphatic System

One of the key elements in understanding how our immune systems function comes from an appreciation of how it interfaces with both the bloodstream and the lymphatic system (Fig. 2.1). Blood consists of red blood cells, which carry oxygen and give it its red color, and a smaller component of white blood cells, which are part of the immune system and help us fight infections. The need for a *lymphatic system* arises from a major plumbing problem posed by our bloodstreams. Blood leaves the heart in large arteries, headed for the tissues to deliver oxygen and an assortment of food substances, and to pick up carbon dioxide and other metabolic waste products. As arteries leave the heart, they branch into smaller and smaller blood vessels, and finally into the smallest blood vessel structures called *capillaries*. Capillaries are found everywhere in the body. This is the point in the blood circulation where food and oxygen leave the blood circulation and enter the tissues.

Both oxygen and food products are ultimately dissolved in the fluid component of blood, and so delivery of these materials to the tissues via the capillaries results in seepage of some of the blood fluids (but not blood cells) out of the blood circulatory system. This indeed poses a problem: What happens to all of the fluid leaking into the tissues? Where can it go? It can't get back

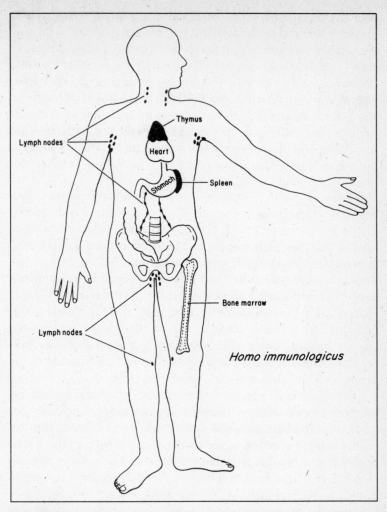

FIGURE 2.1. *Homo immunologicus.* The immune system consists of a number of organs and tissues scattered throughout the body. Each of these organs and tissues is composed of immune system cells, such as lymphocytes (T and B cells) and macrophages. The cells circulate between organs using both the blood and lymph pathways.

into the bloodstream, for the bloodstream is under hydrostatic pressure provided by the constant pumping of the heart. The only reasonable solution is to provide a proper drainage system to collect these fluids—called *lymph*—and return them to the bloodstream. This is the job of the lymphatic system, composed of numerous *lymphatic vessels* that collect these fluids throughout the body and return them to the blood circulation in the great veins of the neck (Fig. 2.2). Everywhere in the body that there are capillaries (and that is *everywhere*), there are lymphatics to drain away spilt fluids. This is one of the most important, yet delicate, operations in the body.

The primary physiological reason for a lymphatic system is thus to make circulation of the blood a completely closed system. Blood leaves the heart and travels through ever-smaller arteries until it reaches the arterioles and then the capillaries. The majority of the fluid portion of the blood continues on through the capillaries and back into the venules and veins, from whence it returns to the heart. This part of the blood circulation is already "closed." However, some of the *fluid* portion of the blood escapes into the surrounding tissue in the capillary beds, making the system open or "leaky." The lymphatic system picks up this spillage and returns it to the blood circulation just upstream from the heart, making closure of the system complete.

Mammals have cleverly co-opted the lymphatic system for another purpose directly related to immunological defense. Because the lymphatic system drains literally every cubic millimeter of the body, inspection of the contents of the lymph is an excellent way to detect the presence of foreign and potentially harmful substances that may have breached the body's natural barriers to infection, such as the skin. What mammals have done is to place clusters of immune tissue at "inspection stations" along this route to examine the contents of the lymph in minute detail. These inspection stations are called *lymph nodes*. Lymph node "filters" are scattered throughout the body, although most are clustered in central areas such as the deep chest, the groin, the axillae, and a few other regions (Fig. 2.1).

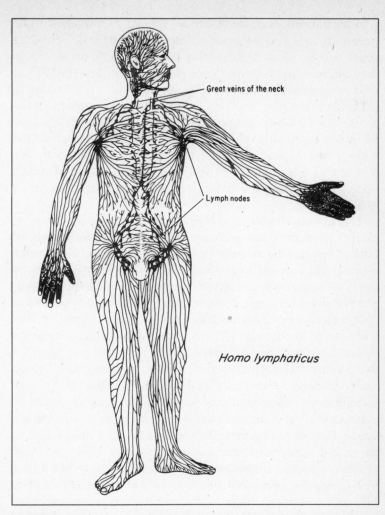

Great veins of the neck

Lymph nodes

Homo lymphaticus

FIGURE 2.2. *Homo lymphaticus.* Essentially every cubic millimeter of the body is drained by the lymphatic system. The lymph fluid collected from body tissues passes through lymph node "filters"; cells inside the lymph nodes examine the lymph for foreign matter. The existence of any biological material in lymph that is not "self" triggers an immune reaction that may ultimately involve the entire immune system. After passing through the lymph nodes, the lymph fluid, which is fluid spilled from blood vessels, drains back into the blood system at the great veins of the neck, such as the jugular vein.

Lymph enters each node via an *afferent lymph vessel*. The lymph, as well as whatever it has brought in with it, percolates down through the various layers of the lymph node before it exits through an *efferent lymph vessel*. Inside the node, arranged in various layers and compartments, are the cells needed to make antibodies. There are really only three cells that we need to know about: macrophages, T cells, and B cells. These three cells are found among the white blood cells that circulate throughout the body in the bloodstream, and they are also found inside the lymph nodes and spleen. (The spleen is itself a kind of lymph node.) We shall see how these three cells work together in just a moment.

This extensive network of lymph vessels can also serve a more ominous function. It is a convenient travel route for individual tumor cells that break away from a primary tumor mass in the process called tumor *metastasis*. Tumors, at least in the early stages of their growth, are served by both blood and lymph vessels, just like any other tissue. Escaped tumor cells that are swept up in lymph fluid drained from a tumor site are often trapped (at least temporarily) in the next lymph node "downstream" from the tumor. Thus with many kinds of cancer, it is routine for surgeons to remove, and submit to pathologists for inspection, lymph nodes draining a tumor they plan to remove or treat by drugs or radiation. The absence of cancer cells in the nodes is usually an encouraging sign, and may allow the physician to select a less drastic form of postoperative follow-up treatment than might be the case if tumor cells were present.

So how does an antibody response begin, and how does it end? We can see this by following the sequence of events set in motion by a minor accident. Imagine that you fall and scrape your knee on a gravelly surface. The skin breaks open, and germs, dirt, and other substances gain entrance to the underlying tissues. These areas, like every other part of the body, have lymphatic drainage for the reasons we have already discussed. Material entering through the skin abrasion is swept into the lymph fluid and carried downstream through a series of coalescing lymphatic vessels toward the site where the lymph rejoins the blood. But along the way, as we have seen, are a series of filters—inspection stations where everything in the lymph is dissected, probed, and analyzed—the lymph nodes.

The first line of defense in these lymph nodes is a type of cell called a *macrophage*. The word macrophage derives from the Greek meaning "big eater." They are basically the vacuum cleaners and filter feeders of the immune system. They ingest and digest (and often regurgitate) everything in sight that is not part of a normal healthy tissue. Nothing gets by them. Macrophages eat everything that manages to penetrate the body's outer perimeters, and they eat cells in the body that get old and die. They are completely indiscriminate in their tastes.

While this function of the macrophage is useful in keeping the place clean, it also plays an important role in the initiation of immune responses. It slows down everything passing through a lymph node long enough for it to be inspected, and for a decision to be made about whether or not it represents a threat. If among the foreign matter entering the lymph node there should chance to be a bacterium, or a potentially pathogenic virus, a macrophage will duly eat it and regurgitate some of the fragments of the invader onto its surface. It is the job of the *T cells* to check the surface of the macrophages constantly to see what they have been eating. If the T cell spots something foreign dribbling down the face of the macrophage, it will realize that a potential pathogen may have breached the outer defenses somewhere in the body and made its way into the lymph node. One type of T cell, called a *helper T cell*, will then seek out an appropriate *B cell* in the lymph node to initiate production of antibodies. The T cells that have been activated by an antigen on a macrophage release chemicals called *lymphokines*, which are needed by the B cell to start producing antibodies. Using the lymphokines provided by the T cell, the B cell will mature into a *plasma cell*, which is essentially a factory able to make enormous quantities of bacteria-fighting antibody molecules. The antibodies are released into the lymph fluid leaving the lymph node and eventually make their way to the bloodstream when the lymph and bloodstreams merge in the great veins of the neck.

The antibody diffuses throughout the body's entire blood vascular system and sticks to any copy of the foreign invader found in the bloodstream. And now the macrophages come back into play. There are also macrophages found in the bloodstream, and indeed all throughout the body. Macrophages have special receptors that allow them to scoop up anything that has an antibody attached to it. This is an extremely rapid and efficient way of reducing and clearing nearly any microbial infection. Once the invaders are cleared from the periphery of the body, and stop migrating into lymph nodes, the antibody response simply shuts down.

The killer T-cell arm of the immune response comes into play especially during viral infections. The important thing that distinguishes a virus from a bacterium in terms of causing disease is that the virus, in order to function at all, has to penetrate inside a living cell. Bacteria can cause disease just by floating around free in the bloodstream. But viruses are not really living cells, like bacteria. They need to get inside a healthy cell and take over its machinery to reproduce themselves, and that is where disease can come in. Diversion of a cell's resources to supporting viral replication usually means the cell cannot carry out its intended function, and there are no superfluous cellular functions in the body.

If the virus makes it to one of the lymph nodes, it will be eaten by macrophages just like everything else. Again, the helper T cells will be alerted by the presence of viral proteins on the macrophage surface. And they will again cooperate with B cells to produce antibodies against the virus. But now the second arm of the immune system is also brought into play. The CD4 helper T cells also help the CD8 killer T cells to mature into their fully active form. The mature killer cells are capable of directly attacking and destroying any cell in the body that has been infected by the invading virus. Note that in this case the immune system is not attacking the pathogen directly. Rather, it is eliminating the "self" cell in which the pathogen is hiding. As we shall see, this process harbors the seed of a major problem; it is getting dangerously close to autoimmunity.

After a brief period of interaction in the lymph node, both the helper and killer T cells slip out into the lymph stream and make their way from there into the blood. They patrol around the body through the arteries and veins, seeking sites where the virus they encountered in the lymph node may have infected other cells. Because the infected cells will lie outside the bloodstream, the T-cell partners must cross out of the bloodstream and into the surrounding tissue. They usually do this near a capillary bed. Following chemical signals released by infected cells, they migrate to and gather at the site of the infection and begin attacking any cell showing signs of having been violated or compromised by an internal (intracellular) pathogen. The killer T cells begin a slow but steady process of search and destroy, egged on by the helper cells who provide them with chemical stimulants that aid them in their mission of destruction. The helper T cells also release chemicals that attract macrophages to the site and then release still more chemicals that stimulate the macrophages into a veritable feeding frenzy. The macrophages help the killer T cells in their mission of destruction, and then clean up the carnage afterward. The 1-2-3 punch delivered by the helper–killer–macrophage trio is an important part of a key immune defense mechanism called *inflammation*.

The Bone Marrow and the Thymus

Before we leave this short introduction to the immune system and how it functions, we should have a brief look at two of the "master organs" of the immune system: the bone marrow and the thymus.

The cardinal feature of bone marrow that resulted in its apotheosis to master organ status is that it is the site in the body where all of the cell types of the blood and immune system are made. Bone marrow is where the blood *stem cells* reside. Stem cells are unusual: They are primitive, undifferentiated cells; they look like no cell type in particular, and have no real function except to give rise to other cells. But stem cells give rise to other cells in a unique way. Most cells divide and produce two "daughter" cells that are exactly like each other, and like the mother cell. Normal cell division is thus said to be *symmetric*. Stem cells undergo *asymmetric* division. One daughter is an exact replica of the mother cell, that is, another stem cell. But the other daughter is slightly more *differentiated* (proceeding toward a fully mature form) along some pathway that will eventually lead to a functional and highly specialized cell type like T cells or B cells. The stem cell is thus self-renewing, and at the same time it produces daughters that go on to make something of themselves. Bone marrow stem cells give rise to all of the mature cells, both red and white, that circulate in the blood. All white cells produced by the bone marrow, whether found in blood or in tissues, are part of the immune system.

Because there is a tremendous amount of cell division going on in the bone marrow—more than almost any other place in the body—the bone marrow is particularly sensitive to drugs, chemicals, and radiation used to treat tumor cells. Most anticancer drugs used in cancer *chemotherapy* take advantage of the fact that tumor cells are in almost continuous cell division. Rapidly dividing cells are also especially sensitive to radiation. Thus the sensitivity of the marrow to these treatments (the *myelotoxicity* of the treatments) is often a major limiting factor in their use for cancer therapy. During radiation treatment, great care is taken to shield bones from the radiation beam so as to protect the marrow. In chemotherapy, because the drugs used for treatment circulate freely throughout the body, not much can be done to protect the marrow, so the amount of the drug given must be tailored to what the marrow can tolerate, rather than what is needed to eradicate the tumor. This is a constant source of frustration to medical oncologists (physicians who treat cancer with drugs, as opposed to surgery or radiation).

If animals are given intensive whole-body radiation, the bone marrow is destroyed and the animals usually die in a week to ten days, either from anemia (lack of red blood cells to carry oxygen) or from a lack of white blood cells to fight infections. High doses of radiation kill mature white cells circulating in the blood, as well as the bone marrow stem cells and the intermediate stages of blood cells developing in the marrow. Circulating red cells are not killed by the radiation, but they normally live only a short time anyway. Because the bone marrow stem cells giving rise to red cells are killed by radiation, the red cells are not replaced as they die naturally in the circulation. Hence, withing a short time of receiving high levels of radiation, the blood system and marrow are completely emptied out, leaving no immune system.

If the radiated animals are given a compatible *bone marrow transplant* soon after they are irradiated, they can be completely rescued from the lethal effects of the radiation. Although their own red and white blood cells, and marrow stem cells, are completely destroyed by the radiation, the donor bone marrow will repopulate the recipient's empty bone marrow spaces and begin to function perfectly normally. After a few months, the recipient will have a complete set of red cells, plus the white cells needed to make an immune system—all of donor origin. This is a major piece of evidence that stem cells for *all* cells of the blood reside in the bone marrow. Because the blood cells in an animal receiving a bone marrow transplant are from a different genetic origin than the rest of the animal, the animal is called a *chimera*. Bone marrow transplants are used in a number of situations in humans to treat disease; those who survive (it is a very risky procedure, and thus not used lightly) are also genetic chimeras for life. We will talk about bone marrow transplants in humans at several points later in this book.

The *thymus* is considered by many to be another "master organ" of the immune system. Not bad for a clump of tissue whose function was still unknown barely three decades ago! Its name, interestingly, was originally taken from the thyme plant, because some early anatomist with a botanical bent thought it resembled the shape of the thyme leaf. The thymus, by the way, is not discussed only in medical textbooks; it shows up on restaurant menus as sweetbreads!

Long before its role in the immune system was known, the thymus was recognized as being *lymphoid* in character—that is, filled with cells of the lymphocyte series. In children and young adults, the thymus is a relatively large, robust organ that lies just above the heart (Fig. 2.1). However, as we age, the thymus gradually atrophies, becoming more fatty and fibrous and less lymphoid. Because of this pattern of atrophy, and the presumed lack of any useful function, the thymus was often heavily irradiated or surgically removed secondary to some other procedure that was being performed. In one famous study in Chicago, children to whom this was done developed a number of immunological abnormalities later in life, including an increase in susceptibility to cancer. The practice was then immediately discontinued.

We now know that the thymus is the site of maturation of the subset of lymphocytes known as *T lymphocytes*. The *T* in fact stands for "thymus-influenced." Several investigators in the early 1960s found that if the thymus is removed in animals at the time of birth, they display major immune deficits later in life. They cannot overcome viral infections or reject tissue transplants, they have an increased incidence of many cancers, and they have trouble making certain kinds of antibodies. Removal of the thymus *later* in life has little effect on immune function.

Cells that are clearly lymphocytes, but which do not yet have any of the markings of T cells, arrive at the thymus from the bone marrow. These "pre-T cells" arrive via the bloodstream and complete their maturation in the thymus. During this maturation period, two critical things happen. First, the T cells decide to become either helper T cells or killer T cells. Second, they acquire one of the principal hallmarks of the immune system: the ability to distinguish self from nonself. This is one of the most important features of the immune system and is what earns the thymus status as a master organ. It is absolutely critical in organ transplantation, which we shall look at a little later on.

Before closing, let us note two important points about the immune responses we have just described that will be important in understanding the rest of this book. First, immune responses are exceedingly powerful. If not controlled in the most careful and precise manner—if they should ever get out of control—they can cause very serious damage. As we shall see, this in fact does happen, and is itself an increasingly important contributing factor in human disease. The second point is that the helper T cell stands at the top of a pyramid of immune reactivity, aiding and abetting and directing nearly every facet of the immune system in its efforts to protect us from damage by invading pathogens. Virtually every immune response we are able to mount is almost totally dependent on the helper T cell. Tragically, it is the helper cell that is itself the target of, and is completely destroyed by, the AIDS virus. And so, long before we get to the chapter on AIDS, you will come to understand why this is such a profoundly devastating human malady.

THREE

Living in the Bubble:
Primary Immune Deficiency
Diseases

So now we know a bit about how the immune system works. We know the key players, and we have a feeling for the strategy and to some extent the tactics used in the game. There is no question that the immune system was a positive force during human evolution, a wall that stood between us and a world of largely invisible predators. For most of our time on this earth as a species, *Homo sapiens* has lived in daily, intimate contact with potentially pathogenic predators that pervaded the soil we trod, tilled, or lived on; infected the animals we tended; and covered the plants we ate. The diseases caused by these microbes were perhaps *the* major reason for a life expectancy that was, a dozen generations ago, less than half what it is today.

But what about today? We don't live in the mud anymore. In modern industrial societies, with strictly enforced public health and hygiene codes, can there be that many pathogens in our environment? Are we still so dependent on an immune system? Not an unreasonable question to ask. To find the answer we do not need to carry out complicated experiments in the laboratory. We need only turn to one of nature's own experiments to see what the consequences would be; we need only look at human *primary immune deficiency diseases*.

Immune deficiency diseases are, when we think of it, pretty much to be expected. The immune system is a rather large collection of different cells and molecules, each with a specific and important function in the defense against microbial infection. Like any complex system—human, mechanical, or electronic—different parts of the system can stop working at different times and under different conditions. Each of the components of the immune system has been observed to break down at one time or another, and these breakdowns provide an unambiguous answer to the question just raised: The loss of the smallest component of the immune system can be disastrous. On the other hand, these tragic experiments of nature have often provided critical insights into how the immune system works, as we shall see.

The possibility that disease states could result from a breakdown in immune components must surely have been on everyone's mind from the earliest days of immunology, but the first medical report linking a specific clinical problem with a specific immune defect did not appear until 1952. For the next twenty years or so after this initial report, however, there was an incredibly exciting burst of interplay between physicians, mostly pediatricians, and basic research scientists studying the immune system in animal models. The result of these interactions was not just a rapid expansion of our ability to identify and treat a whole new range of clinical disorders. A careful analysis of immune deficiency diseases in humans greatly extended our understanding of how the immune system itself is designed. The most profound result was this: Together with studies of immune deficiency states in animals, as discussed earlier, these diseases would reveal to us that we have not one, but two immune systems. We would find ourselves endowed with B cells and antibodies on the one hand, and T cells with their lymphokines on the other. In the space of a few short years, some of the most confusing aspects of immunity would become clear.

We begin with a bit of context. In humans, there are two distinct categories of immune deficiency disorders, which are distinguished by the nature of the origin of the disease. Innate or *primary immune deficiencies* are genetic in origin. That means that the defect is in a gene inherited from the child's parents. The defect is therefore present (although not necessarily obvious) at birth. The majority of these disorders manifest themselves within the first few years of life, and some are apparent at or shortly after birth. Primary immune deficiencies are thus the province of pediatricians, and in fact represent a distinct subdiscipline within the field, although all pediatricians are trained to detect such problems. There are over seventy well-delineated primary immune deficiencies in humans. The overall frequency of primary immune deficiencies is about one in ten thousand; with approximately four million births each year in the United States, we would thus expect to see somewhere around four hundred new cases annually.

Secondary or *acquired immune deficiencies* are not the result of inherited genetic abnormalities, but arise secondary to some other disease process, or after exposure to drugs or chemicals that are toxic to the immune system. These are by far the majority of immune deficiencies seen clinically, and are certainly *not* the exclusive province of pediatricians. Common causes of acquired immune deficiencies are malnutrition, stress, burns, certain autoimmune disorders, and certain viruses. Patients receiving organ transplants are deliberately immunosuppressed to facilitate acceptance of the transplant, and may thus be immune deficient—even lethally so. Many drugs used to treat cancer patients are also potent immunosuppressants. These latter two groups may well comprise the largest category of patients with acquired immune deficiency. The most prominent example currently of a secondary immune deficiency, however, is AIDS—the Acquired Immune Deficiency Syndrome—which is secondary to infection with what has become known as the Human Immunodeficiency Virus, or HIV.

In this chapter we will have a look at a few of the primary immune deficiencies in humans. AIDS, which has its own story to tell about the strengths and failings of the immune system, will be dealt with separately in a later chapter. Although as noted there are primary deficiency states affecting virtually every immune compartment, it will be sufficient to look at just those that affect the T- and B-cell compartments of the immune system to get some indication of how profound these diseases are, both clinically and scientifically.

A Light Goes On

The first disease to be thought of as a specific defect in an immune system compartment was reported by Dr. Ogden Bruton in 1952. Bruton was at the time a colonel in the United States Army Medical Corps. Although a career military officer, he was also a general practitioner catering to the needs of soldiers and their families. He was called in to consult on the case of a military dependent, a young male patient, aged eight, who had had repeated bouts of bacterial infections, of increasing frequency and severity, from about the age of four onward. Each episode was accompanied by various combinations of fever, vomiting, joint pain, and a variety of other symptoms requiring hospitalization. He would respond well to penicillin and sulfa drugs, and usually would be discharged after a short period, only to return a few months later with the same symptoms. Cultures of various body fluids suggested pneumococcal bacteria as the most frequently seen infectious agent, so an attempt was made to immunize him with a vaccine made from killed pneumococcal organisms. After repeated vaccinations, the doctors were unable to find any sign of pneumococcal antibodies in his serum. He was then injected with a series of other bacterial vaccines, with the same result: No antibodies to any injected bacterium could be demonstrated in his serum.

Finally, Bruton and his colleagues decided to have a look at the child's blood proteins using a new technique that allowed visualization of each of the major protein groups in blood. It had been established some years earlier that antibodies belong to a protein group called *gamma globulins*. Gamma globulin is readily detectable in the blood of healthy individuals, and it is especially prominent in the blood of persons fighting off a bacterial infection. In Bruton's young patient, gamma globulin was completely absent, even in the midst of an ongoing infection. In an attempt to treat one of his episodes of bacterial infection, the youngster was given an injection of pooled gamma globulin harvested from the blood of human donors who had, simply in the course of normal daily life, experienced and overcome a variety of infections. Such gamma globulin fractions contain an assortment of antibodies against a wide range of common environmental pathogens. The passively administered gamma globulin worked very well, although as expected it wore off after four to six weeks. In the four years immediately preceding this report, the patient had experienced nineteen serious episodes of bacterial infection. With monthly injections of pooled human gamma globulin, he had had fourteen months without any problems (and in fact remained so for many years).

The disease state typified by this young patient is still sometimes referred to as *Bruton's agammaglobulinemia* or *Bruton-type agammaglobulinemia*. The original paper is intriguing because one can almost see the light go on when, after repeated attempts to interpret the disease from the point of view of the *pathogen*, it suddenly dawned on Bruton and his colleagues that the problem might be on an entirely different level—that of the patient himself. Instead of looking at how the infectious agent might have changed so that it could escape immune destruction, they asked whether something might be wrong with the patient's immune system, such that a relatively innocuous bacterium could now wreak destructive havoc. This was a completely new way of approaching problems of infectious disease. In hindsight it seems trivial, almost obvious, and compelling. But these subtle mindshifts are often amazingly difficult births!

Given the numbers of young patients treated for immune deficiencies of this type today, and the fatal outcome if not treated correctly, one can only speculate how many immunodeficient children must have died of bacterial disease before 1952. On the other hand, these youngsters have normal immunity to all but a very few viruses. This was, as we shall see, one of the first clues to the dual nature of the immune system. It is possible to lose entirely one's immune defenses against bacteria (antibodies) without being seriously impaired in the response to viruses (T cells). The treatment recommended by Bruton in his original case is still the standard today: aggressive antibiotic therapy to clear up any ongoing infection, followed by administration of pooled human gamma globulin to prevent further occurrence. Bruton described the progress of his first patient ten years later. The boy went on to a perfectly normal childhood and adolescence, finished high school with honors, and entered college. Although he never gained the ability to make his own antibodies to the bacteria that threatened him, his health was comparable to others in his family at all times.

One striking characteristic of Bruton's agammaglobulinemia is that it affects almost exclusively young males. That is because the gene causing it is located on the X "sex chromosome," and the disease is thus said to be *X-linked*. (The generic term for deficiencies of the type described by Bruton is now *XLA*, for *X-l*inked *a*gammaglobulinemia.) Females have two X chromosomes; in order for a baby girl to develop XLA, she would have to inherit two defective Bruton genes, one on each chromosome. The likelihood of this is very low. The father would have to be suffering from XLA, and the mother would have to contribute a defective gene. Males have one X and one Y sex chromosome. If the single X chromosome has a defective Bruton's gene, then that's it—there is no second X chromosome around to rescue the situation.

Bruton speculated in his paper about whether this child's problem stemmed from a congenital (primary) defect or was secondary to some other disease or injury (acquired). Because his patient's problems were not particularly serious before age four, Bruton guessed the underlying disorder must be a secondary immune deficiency. He reasoned that if the condition had been present since birth, the associated problems would have manifested sooner. Although incorrect, this was not an unreasonable conclusion at the time. Bruton could not have known that his patient had a defect restricted to his B cells, the cells that make antibody.

The B cells would not be defined for another twenty years. (In his ten-year retrospective on his first patient, Bruton acknowledged that subsequent events confirmed the congenital nature of the disorder.) While B cells are certainly important in the defense against bacteria, the body has a variety of other defense mechanisms as well. The new infant is equipped with antibodies that cross the placenta during pregnancy, and these can be supplemented with antibodies in breast milk after birth. Together these inherited antibodies can provide protection during much of the first year of life. In addition, macrophages are able to provide a primitive sort of protection against bacteria. With good hygiene to minimize levels of bacteria in the environment, it is entirely possible for a child with XLA to remain disease free for several years.

At the time of Bruton's report, knowledge of the immune system was still too primitive to allow an analysis of just where the defect in this patient might lie. But once reported, suddenly everyone seemed to realize he or she had been seeing these sorts of diseases all along. A new problem was at hand for both physicians and scientists to get their teeth into. However, before we look at what they found out about this particular defect, let us take a look at another immune deficiency in children described some thirteen years later—the *DiGeorge syndrome*.

Dr. Angelo DiGeorge first described the syndrome that would bear his name at a meeting of the Society for Pediatric Research in 1965. DiGeorge had been interested for some time in infants born without a thymus gland. This congenital condition was usually part of a larger syndrome in which the parathyroid glands are also absent, and in which there are characteristic abnormalities in head and neck structures that allow these infants to be identified shortly after birth by an alert pediatrician. All of these structures—the thymus, the parathyroids, the affected head and neck elements—have a common embryological origin, so it is clear that the underlying defect in this syndrome is an anatomical abnormality during fetal development that leads to malformed structures, including the thymus. The parathyroid dysfunction causes problems with calcium metabolism, which in turn leads to cardiac problems and mental retardation. The overall condition is spotted quickly after birth because of this calcium defect, which also leads to muscle dysfunction and seizures. Almost secondarily, it had been observed that these children are unusually susceptible to fungal and viral infections. Thus completely aside from the absence of a thymus, these children have severe physiological problems and rarely survive beyond a few years of life.

In 1965, in fact, the function of the thymus in humans was still unknown. Based on studies then in progress in animal models, which we will discuss shortly, DiGeorge suspected that athymic infants might be immunodeficient. Thus it was that when, only a few months before the Pediatric Research meeting, a nine-month-old infant with all the symptoms of what we now call the DiGeorge syndrome was brought to him, DiGeorge carried out a limited number of tests for immune function. What he found was that the infant was unable to reject a small piece of transplanted skin. Skin graft rejections were known by this time to be caused by some sort of immune cell, rather than antibody, but the exact nature of this cell was not yet known. On the other hand, antibody levels in this infant were essentially normal. In the time since Bruton had first described his agammaglobulinemia syndrome, other youngsters with the same disorder had been more extensively evaluated for immune function and were found to have perfectly normal graft rejection capability.

Although many other primary immune disorders would eventually be described, a surprising number of them seemed to fit into either a general Bruton's category (loss of antibody function, but graft rejection normal), or a generalized DiGeorge-like category (antibody function intact, but loss of the ability to reject grafts). Thus the clinical evidence was moving rapidly toward a definition of two separable and distinct immune functions in humans, with fundamentally different underlying modes of action. Exactly the same conclusions were being reached by scientists studying immune responses in laboratory animals, where the important parameters could be more definitively manipulated.

When the two approaches merged, the result was a veritable explosion in information about how the immune system is put together.

We saw earlier how studies by Bruce Glick and his colleagues provided an important clue that the immune system might actually be composed of two separate arms. Until that time, antibodies were the only known immune mechanism. Their discovery that the bursa in chickens controlled the ability to make antibodies and fight bacterial infections, but seemed to have no effect on the ability to reject grafts or control viral infections, suggested the existence of a second immune mechanism dedicated to the latter two functions. But what that second mechanism might be was completely unknown.

As yet unaware of the findings of Glick and his associates, Dr. Robert Good, then a young resident at the University of Minnesota, embarked on a series of studies the following year that would eventually lead to a remarkably synergistic interaction between the study of animals and humans, and assure Glick (as well as Good) a place in the history of immunology. A peculiar clinical case of a man with simultaneous agammaglobulinemia and benign thymoma (tumor of the thymus) triggered Good's interest in the possible role of the thymus in the immune response. He carried out a series of experiments on rabbits and mice in which he removed the thymus and then observed the animals for immune defects. He found none, and almost gave up the study. But then a fellow resident ran across Glick's paper and showed it to Good. What this resident might have been doing reading *Poultry Science* we may never know, but it provided one of those fortunate coincidences that changes life in ways we could never anticipate. Good immediately recognized that perhaps he hadn't been wrong after all. He rushed back to the lab, and this time removed the thymus from very young animals, within a day or two of birth. The results were quite different this time. The thymectomized animals showed good responses to bacterial antigens, and had nearly normal levels of serum antibodies, but they were completely deficient in the ability to reject skin grafts. They were also highly susceptible to viral infections.

Unfortunately, mammals do not have a bursa of Fabricius, so researchers went back to the chicken, which has both a bursa and a thymus. In the chicken, it is possible to remove the thymus and bursa independently and thus clearly assess the role of each. The results of these experiments were unambiguous, allowing formulation of a new model for how the immune system functions. The new description accounted beautifully both for the results of animal studies by many groups around the world and for the emerging evidence from primary immune deficiencies in children. It was proposed that the thymus and bursa control two separate and distinct compartments in the immune system. The bursa (bone marrow in mammals) controls the development of cells (B cells) that are responsible for producing antibodies. The thymus controls the development of cells (T cells) that are involved in what are now called *cell-mediated immune responses:* principally graft rejection and viral defenses. Removal or inactivation of these organs close to the time of birth results in immune defects that are operationally indistinguishable from the clinical problems exemplified by DiGeorge-type defects and Bruton-type agammaglobulinemia. This proposal, which was rapidly accepted by basic researchers and clinicians alike, profoundly changed immunology for all time. It was now apparent that immunologists would have to contend with (and explain) *two* immune systems, rather than just one.

Going Naked in the World: SCID

A few unfortunate children (about one in one hundred thousand) are born with defects in both the T-cell and B-cell immune compartments. The results are almost always devastating. The most deadly forms of this disorder are grouped together under the name *severe combined immunodeficiency disease (SCID)*, characterized by overwhelming infections by almost every microbe known—bacteria, viruses, fungi, and parasites. Without a doubt, these infants are, for all practical purposes, immunologically identical to AIDS patients in the end stages of their disease.

In most forms of SCID, the collapse of B-cell function is secondary to an absence of T cells. Without T-cell help, B cells are unable to produce antibodies. Thus unlike Bruton's XLA, where B cells are essentially completely absent, in most forms of SCID (including X-linked SCID) there are considerable numbers of B cells present. The T-cell deficiency usually shows up first, in the form of susceptibility to fungal and viral infections. Protection against bacterial disease is conferred during the first year by antibodies crossing the placenta from mother to child, and in breast milk. Infants who escape or somehow survive the viral and fungal episodes will eventually also be assaulted by round after round of bacterial infections.

Keeping these children alive is a daunting task, for which only the best hospitals are equipped. Even then, many youngsters do not survive. Too often they are brought to the hospital with advanced microbial infections that simply cannot be controlled in time to prevent death. Once SCID became generally recognized, and management of it made a standard part of a pediatrician's training, it was realized that many of these infants even developed fatal complications from immunizations with highly (but often not completely) attenuated vaccines shortly after birth. How many of these infants died from vaccinations prior to this realization is unknown; fortunately, SCID is a rather rare condition.

As can be imagined, the outlook for infants with SCID is extremely bleak; with the very best management and supportive care, they may survive the first two or three years of life, rarely beyond, unless they receive a bone marrow transplant. Initial bacterial and fungal infections in these infants can be managed by antibiotic treatment, but it is virtually impossible to manage viral infections, and these are a common cause of death. The only hope for long-term survival is a bone marrow transplant, preferably from a closely tissue-matched sibling. Bone marrow from a healthy donor, with its stem cells, should be able to replenish completely the cells of the immune system, including both T and B cells. This technique was pioneered by Dr. Robert Good himself, among others. Even then the outlook is poor; only about a third of those so treated survive beyond a few years.

The best-known case of a "SCID kid" was a young boy named David, who became known to millions of Americans and others around the world as the "Bubble Boy." David was born in 1971. Because a previous male child born to his parents had died from SCID a few months after birth, the risk of a second SCID child was known well in advance. The form of the disease in this family was X-linked, and since the father (being healthy) was obviously not a carrier, it was clear that one of the mother's X chromosomes carried the defective SCID gene. Therefore half the daughters produced from this marriage should be carriers also, and half completely normal. They had in fact already produced a normal daughter, Katherine, born three years before David. Among any sons produced, all would inherit the father's Y chromosome; half would inherit the mother's good X chromosome, and half the bad. Thus overall there was a one in four chance that the next child would also have SCID at birth; David's parents decided to take the risk. Amniocentesis at the fifth month of pregnancy showed that the child would be male; the odds were now one in two. At that time there was not yet a way to predict from amniocentesis whether a male fetus has a gene for SCID. That would not have mattered in this case, because the parents did not consider abortion an option.

Thus David was delivered by cesarean section and transferred within seconds to a sterile incubator until his immune status could be determined. It was soon obvious that he carried the defective gene causing SCID. As it turned out, David's physician had had considerable experience with SCID. In fact, he had the remarkable experience of treating fraternal twin boys born with a form of SCID, who, kept in a sterile environment after birth, spontaneously recovered their immune function after two and a half years. There was no way to know at the time that this would be a virtual impossibility with the type of SCID David inherited.

Young David quickly became the longest-living SCID patient, untreated except for sterile isolation. The major rationale for this highly unusual approach to managing a "SCID kid" was the hope that if he could be kept alive long enough by keeping deadly pathogens away from his body, either a suitable bone marrow donor could be found or his immune system might somehow establish itself and allow him to fend for himself. Because virtually nothing was known about the basis of this desease in the early 1970s, and given the example of the twins who had spontaneously recovered, and several examples of successful marrow transplants, these were not unreasonable hopes. David was repeatedly tested for any signs of T- or B-cell responsiveness; there were none. When he reached the toddler stage of development, he was moved into a sterile tent that allowed him to crawl and eventually stand. The tests continued; still no response. When he began to walk and run, the tent became the "bubble," a complex system of interconnecting plastic tubes that allowed considerable freedom of motion, within obvious limits. NASA even built a small spacesuit for him when David was six so he could be taken into the outside world as well. He outgrew it within a year. A sterile transporter was also developed so that he could be taken home and develop a sense of belonging to a family. He was given the basics of an education in his bubble and at home; his nurses and tutors found him a bright, somewhat mischievous youngster, virtually indistinguishable from other boys his age. But his immune system never developed; the only thing between him and almost certian death was a few millimeters of plastic sheeting and high-quality air filters.

As David continued to grow and develop, it became clear that something simply had to be done. He was healthy and vigorous, and at twelve years of age showing the first signs of normal sexual maturation. He had not yet begun showing outward indications of a curiosity about sexual matters, but clearly his situation was approaching a critical stage. No one had really thought this far ahead; no untreated SCID youngster had ever lived this long. His medical team members found themselves in an ethical dilemma of gigantic proportions, with no guidelines whatsoever for how to proceed. The prospect of maintaining him any longer in a sterile bubble—for how long? ten years? twenty? fifty?—was becoming increasingly untenable. How do you talk with a child like this about the future, a concept he now understood only too well? Finally, it was decided to give David a bone marrow transplant, with his sister (then fifteen) as the donor. In most cases, marrow from sibling donors has a higher chance of successful acceptance than marrow from a complete stranger. However, David and his sister were not particularly *histocompatible*, or tissue compatible, which is one reason a bone marrow transplant was not attempted earlier.

Nevertheless, it was decided to proceed. Marrow was removed from David's sister and treated to remove mature T cells. Mature T cells are not found in bone marrow per se; they are a contaminant from the harvesting procedure when blood vessels woven throughout the marrow are broken, allowing mature blood cells to mingle with the precious bone marrow stem cells. It was thought at the time that mature T cells contaminating donor marrow might be responsible for a major barrier to successful transplantation: graft-versus-host (GVH) disease, which can be lethal. In GVH disease, mature donor T cells in the incoming marrow, being fully competent immunologically, regard the new host as a gigantic transplant, which they immediately set about trying to reject. The graft, in effect, is rejecting the recipient. This can be fatal in a quarter to a third of patients receiving a bone marrow transplant, which is why such transplants are carried out only in the most serious situations.

David was brought from home in his sterile transporter unit and placed in his original bubble in the hospital. When his sister's bone marrow was ready, David was taken from his bubble in a sterile transporter to a sterile operating room and infused with the marrow. He actually assisted the physicians and nurses in the procedure, which does not require anesthesia. He was kept in a sterile postoperative recovery room and then returned to his bubble. For the next several weeks, everything seemed to go well. He was even allowed to go home for the winter holidays to spend two weeks with his family. But after his return, he developed symptoms that seemed possibly related to GVH disease: weight loss, gradually increasing fever, vomiting and diarrhea, abdominal tenderness. Appropriate steps to control GVH were immediately undertaken, but David did not respond. There were also signs of a viral infection. His condition grew rapidly worse; he finally died on the 124th day posttransplant, in February 1984. David was twelve years old.

At autopsy, physicians and immunologists following David's case encountered totally unexpected results. First, there were no signs of GVH disease caused by his sister's T cells; in fact, there was no sign whatever of any cells related to his sister's bone marrow. The graft had completely failed to take. Since that time (1984), there has been a great deal of concern that either the procedures for removing mature T cells from donor marrow may damage it, compromising the marrow's ability to repopulate the new host's immune system, or that mature T cells, at least in small amounts, may actually help the new bone marrow engraft into the recipient. A little GVH may be a good thing, as one researcher put it. Thus the procedure for reducing the likelihood of lethal GVH may also reduce the likelihood of successful marrow engraftment. Once again we find ourselves walking an immunological tightrope without a net. A slight tip to one side or the other can spell disaster.

The second major surprise at autopsy was the actual cause of David's death, which turned out to be congestive heart failure secondary to B-cell lymphoma. The lymphoma had spread throughout David's system—brain, intestines, lungs, liver. And the lymphoma had arisen from his own B cells, not his sister's. In David's SCID, there are near normal levels of what appear to be normal, mature B cells. They cannot produce antibody (for reasons we will discuss later), but they can serve as targets for potentially cancer-causing viruses such as the Epstein-Barr virus (EBV), which is a common virus present in perhaps 80 percent of healthy humans. It is usually kept under control by the body's T cells. Even when it breaks free it usually causes nothing more serious than the flu-like condition known as *mononucleosis* ("kissing disease"). On the other hand, in T-cell–compromised individuals, such as those with primary T-cell immune deficiencies, immunosuppressed transplant recipients, or AIDS patients, EBV-infected B cells may begin to grow in an uncontrolled fashion. The best guess is that David's sister was an otherwise healthy EBV carrier; her marrow (or the contaminating mature blood) was apparently infected with EBV. Once transferred to David, her marrow did not survive, and the EBV escaped into David's system, where it encountered some of his own B cells. As often happens in leukemias and some lymphomas, David's internal body spaces began to fill with fluid. When this fluid seeps into the cardiac cavity in the chest, it interferes with normal function of the heart, and congestive heart failure is the result.

As can be imagined, David's death was devastating to everyone who knew him—his family, his friends, the medical team that had cared for him. Many of these people required psychological counseling for some time after his death. As with all children studied for immune deficiencies, we learned enormously from David's experience, and what we learned will directly benefit other SCID children in the future. The question of whether David's treatment was appropriate was debated vigorously after his death, and will be for many years to come. This is surely a good thing, for we have few landmarks to guide us in situations like David's. Some felt strongly that using a human life for what amounted to a medical experiment could never be compatible with the dignity of human life. One clergyman declared that

> David's very existence presents the specter of a virtually autonomous medical technocracy at work in our society, a technocracy that is at best only dimly aware of the subtle and delicate boundaries of the human. . . . The creation of David in his bubble seems clear evidence that the medical world, and perhaps Western society at large, has drifted into a kind of technocratic imperialism . . . ; the physicians and scientists who created this project

seem to live rather isolated in their own arcane scientific and technological worlds.

These words were written shortly after David's death. Possibly the person who wrote them softened on some of these points with time. Certainly they were vigorously rejected by David's family, who found great meaning and joy in David's brief life and truly believed that David did as well. The problem with learning from such situations is that they rarely repeat themselves in a directly recognizable form. Certainly there will never be another "Bubble Boy" living a dozen years or more in sterile isolation. There doesn't need to be. We now know this is not a useful approach to treating this tragic condition. That precious information was David's gift to the world.

Gene Therapy: The New Hope

On September 14, 1990, with a slight grimace as a slender needle slid through her skin and into an underlying vein, a four-year-old girl with lovely dark eyes became a part of medical history. Ashanti de Silva became the first human being ever to receive a new treatment called *gene therapy*—the deliberate introduction of a piece of genetically engineered DNA into her cells in an attempt to cure a deadly disease caused by a defective gene.

Ashanti also suffered from a form of SCID. Unlike David's disease, hers was not caused by an X-linked gene, but by an unrelated gene on another chromosome. Both of her parents carried one defective copy of the gene. She was the unlucky one-in-four offspring that would inherit two defective copies—one from each parent. This type of SCID is due to a defective gene for the enzyme adenosine deaminase (ADA). ADA-SCID, as it is called, results in exactly the same defect as David experienced: the inability to mount any sort of immune response at all. It is due principally to the absence of ADA in T cells, which as a result die and are thus unable to carry out any of their critical immune functions, such as helping B cells make antibodies or fighting viral infections.

Other treatments to manage Ashanti's condition had not been successful. The level of her T cells had fallen as low as fifty cells per cubic millimeter of blood—a level ordinarily seen only in patients in the terminal stages of AIDS. Standard medical practice would dictate that the only option left for her would be a bone marrow transplant. But in this particular case there was no well-matched sibling to act as a donor, so a bone marrow transplant would have offered a less than 50 percent chance for success. Fortunately for her, just a few months earlier doctors at the National Institutes of Health (NIH) had finally received permission to carry out the first clinical trials for genetic repair of the gene causing ADA-SCID. Years of laboratory experiments, including repair of the same gene in animals, had convinced both scientists and physicians that the procedure should be both safe and effective in humans. No fewer than seven NIH oversight committees had reviewed the proposed procedure and agreed to it.

Here's how the procedure worked. Ten days earlier, a sample of the young patient's blood had been withdrawn and her T cells (carrying the defective gene) isolated. A copy of a perfectly normal human gene for ADA was introduced into her T cells using what is called a *retroviral vector*. In this case the vector was a retrovirus that ordinarily causes leukemia in a mouse. This might seem like a frightening thing to use in a human patient, but it is completely harmless to humans in the form used. Even in its fully intact form, this virus does not cause leukemia in humans, but for these experiments the virus had also had some of its own genes removed so that it could not reproduce even in a mouse. In place of the virus's missing genes the researchers introduced a healthy gene for human ADA. Although the newly engineered virus had no way of reproducing itself, it was still able to infect a living cell. Once inside the cell, the retrovirus delivered the new gene to the nucleus, where it was incorporated into the cell's DNA and instructed the cell to make normal ADA.

No attempt was made to repair or remove the defective ADA genes from Ashanti's T cells; they were simply left in place, and copies of healthy new genes were added in. Her T cells were then grown in an incubator for a week or two in the presence of T-cell growth factors to increase their number. After that, they were ready to put back into the patient. Because T cells live in the blood, no surgery was needed; they were injected back into the bloodstream with nothing more than a standard needle and syringe.

As expected, there were virtually no side effects. After the injection, the youngster experienced a slight fever for several hours, which then faded. After another injection one month later, the level of healthy T cells in her blood began to climb, and they were shown to be using the newly introduced gene to make ADA. Most important, Ashanti began to show signs of being able to make antibodies, proving that the repaired T cells were working normally. After six injections, her T-cell count was 1,250, well within the normal range for healthy humans. The results were so encouraging that a second child with ADA-SCID was treated by gene therapy just a few months later, with equal success. Both of these young medical pioneers are perfectly healthy, attend school, and are very active. They are involved in, among other things, the national March of Dimes campaign to raise funds for childhood diseases.

To date, these two young patients have each received more than a dozen injections of their own altered T cells. And therein lies the only element of dissatisfaction with the approach taken to treat them. T cells live but a short time in the body, unless triggered by foreign antigen. Eventually, as these children are exposed to more and more environmental antigens, they will build up populations of the longer-lived memory T cells. This process could even be enhanced by planned immunizations, but it is a less than ideal way of reconstituting the immune system. So from the very first the doctors and scientists knew that the ultimate target for their gene therapy would be the *stem cells* that give rise to the T cells. Stem cells, remember, are self-renewing in addition to acting as parents for the production of other cell types. Thus if the defective gene can be replaced in a stem cell, then the procedure has to be done only once, or at most only a few times.

And this, too, has now been done. On May 12, 1993, barely two and a half years after the very first gene therapy, Andrew Gobea was born at Children's Hospital in Los Angeles. A sister born previously had died of ADA-SCID, and Andrew's condition was detected before birth by amniocentesis. Stem cells are found in the bone marrow in adults. They are very rare, accounting for only about one in ten thousand marrow cells. But in humans they are also present in the blood circulation of the fetus, and can thus be harvested from the umbilical cord as it is cut at the moment of birth. Andrew's umbilical blood was collected and rushed to the lab, where the stem cells were isolated and injected with the ADA gene. At four days of age, he was reinjected with his own repaired stem cells. His father put on gown and gloves and actually pushed the syringe plunger that delivered the altered cells. Little Andrew will be monitored closely during his first year of life to see how well the procedure worked. If necessary, some of his bone marrow cells can be removed at a later date for a supplementary repair job. As part of the same clinical plan, several other children with Andrew's disorder have now been treated by the same means.

Theoretically, any of the primary immune deficiencies can be cured by gene therapy, as all are genetically based. There are a few technological restrictions. For the foreseeable future, we will probably be able to treat only diseases caused by a defect in a *single* gene. It is possible that some of the known immune deficiencies are caused by defects in more than one gene; the technical difficulties in correcting genetic diseases goes up dramatically as the amount of DNA involved increases. Fortunately, most of the known primary immune deficiencies appear to involve only a single gene.

A second requirement is that the gene causing the disease has acually been isolated and can be grown in the lab. This is happening with increasing frequency for all sorts of genetic diseases such as cystic fibrosis, muscular dystrophy, or Huntington's disease, in addition to the primary immune deficiencies. We have a long way to go, but new genes are cloned and studied almost every day. And it is in this arena that we come full circle. When David died, his doctors removed some of the lymphocytes from his blood and froze them for future study. Over the past several years, scientists at the National Institutes of Health, working with DNA cloned from David's cells, have succeeded in identifying the defective gene in X-linked SCID. It codes for a protein found on the surface of T cells. The job of this protein is to receive and process a particular chemical signal during the period that the T cell is maturing in the thymus. Without this critical protein—one single molecule out of the thousands that make up a T cell—the T cell cannot complete its maturation in the thymus. The result is X-linked SCID.

The gene for David's disease was reported in a scientific journal in April 1993, just three months after research teams in England and at UCLA simultaneously reported that they had isolated and identified the gene for Bruton's X-linked agammaglobulinemia. Given the pace of genetic engineering technology, a form of all of these genes appropriate for gene therapy may be ready in just a few years. Copies of these genes will also be useful in determining whether fetuses are carrying the disease, leading to better counseling of anxious parents.

Above all else, these "molecular medicine" approaches to treating primary immune deficiency diseases remind us of how incredibly fragile humans are as living organisms. The horrible results of diseases like XLA, David's SCID, and ADA-SCID may be the result of a single nucleotide change—one nucleotide out of thousands that spell out the genetic code for a gene—which in turn is only one of a hundred thousand or so genes that make up a human being. Was this change due to mutation? A chance encounter with a cosmic ray or a toxic chemical? A simple reading error when the DNA was copied from one cell to another? We will never know, but the consequences in terms of the tragedy wrought upon an individual and a family from such a simple event are almost beyond comprehension.

All in all, there are over four thousand diseases in human beings known to be caused by defective genes. Many of these will be amenable to treatment by the type of gene therapy that has been used to treat ADA-SCID. In fact, there are now over fifty approved procedures for gene therapy at selected medical centers across the United States; about thirty of these are already under way. The diseases being treated range from immune deficiencies in children, through brain tumors and lung cancer, to AIDS. Every one of these procedures is the result of years and years of careful work in the laboratory—isolating the gene, studying it, making billions upon billions of copies of it, learning how to control its expression in human cells. Then the clinicians come on board, figuring out the best way to introduce it into a human patient, how to follow the patient once the procedure is completed, what danger signs to look for if there is a problem. As with the first ADA-SCID cases, each procedure is reviewed time and time again by numerous panels consisting of doctors and scientists not involved in the experiments, and by politicians, religious leaders, philosophers, and citizens from the community.

It will be a number of years yet before these procedures are routine enough to be used in the average community hospital, but make no mistake about it—they will be. Some people have expressed concerns about the potential for abuse with these techniques, and rightly so. We are talking, after all, about altering human beings in the most fundamental of ways, by changing their DNA. Strict oversight is an essential part of this procedure, as much as the introduction of the DNA itself. But gene therapy holds enormous potential for correcting diseases caused by defective genes. That potential could not have been more dramatically demonstrated than at a meeting recently between the two young girls treated by gene therapy for ADA-SCID, and David's mother. It was an emotional and touching moment, as much for the doctors and scientists looking on as for David's mother and the two girls.

So, controversial though they may be at present, the techniques of molecular medicine are here to stay. I constantly emphasize to my students that they must study molecular biology now, as undergraduates, for medicine in their time will be profoundly influenced by molecular biology and biotechnology. They will be called upon to make decisions and provide advice about things that even many of their instructors in medical school are only dimly aware of. They will have to grow with the field—read the journals, continue to attend lectures—for many years after they finish their formal training. No one today can predict which diseases will be treatable by gene therapy during their lifetimes. But one thing we can predict—it will be one of the most exciting times in the history of medicine.

FOUR

Hypersensitivity and Allergy

The preceding chapter makes it painfully clear what the absence or failure of certain components of the immune system can mean in terms of human well-being. The immune system truly is a wall that stands between us and the world of microscopic agents of disease. The slightest crack, or missing stone, can spell disaster. So the possibility that this beautiful, elegant defense system could also cause harm was at first very difficult to accept. Some of the very earliest findings pointed in that direction, but were largely ignored. But as scientists looked ever more closely at the immune system during the first half of the twentieth century, that possibility became first a probability, and ultimately a reality. It seemed in a number of situations, at least in the laboratory, that exposure to a foreign antigen did not always lead to protective immunity, but rather to a state in which subsequent exposure to the same antigen could elicit a violent, often harmful, and occasionally fatal syndrome. This phenomenon of overreaction became known as *hypersensitivity*. It would eventually be recognized for exactly what it is: a defense system out of control. But this realization took time.

The Discovery of Hypersensitivity

The story of the discovery of hypersensitivity is, despite the ominous nature of the subject, a rather charming one. Prince Albert I of Monaco (1848–1922) was an amateur scientist with a particular interest in oceanography. Among his philanthropic projects was the outfitting of a handsome 1,400-ton yacht, the *Princesse Alice II*, as a modern, fully equipped seagoing laboratory. Each summer he would invite distinguished scientists to join his on-board scientific staff for cruises around the Mediterranean and out into the Atlantic on various oceanographic missions. (This is a style of doing science that, despite its decidedly nondemocratic overtones, strikes a certain chord in those of us who spend our lives in modern and efficient but architecturally boring indoor university laboratories. Government funding agencies, to which a restoration of this means of scientific investigation has been suggested, have responded rather gracelessly.)

We sometimes forget the quite beneficent influences of enlightened and generous royal patrons. In the summer of 1901, Prince Albert invited a young physiology research assistant from the Sorbonne, Paul Portier, and a distinguished senior scientist, Charles Richet, to join him and his staff as they went looking for specimens of the Portuguese man-of-war (genus *Physalia*). The fruits of the collaboration between Portier and Richet, begun under the aegis of Prince Albert, would lead to a Nobel Prize for one of them (Richet in 1913) and completely alter the way we think about the immune system. But in this, the first summer of the new century, the prince was interested in studying the nature of the toxin used by *Physalia* to stun its prey and, perhaps of more immediate interest to the prince, to injure sailors. The *Princesse Alice II* embarked from Toulon in early July and sailed for about ten weeks. Large numbers of *Physalia* were soon encountered, and in short order Portier and Richet had extracted, concentrated, and begun characterizing the toxin, using a variety of laboratory animals available on the yacht. Their initial work focused simply on the pharmacological aspects of the toxin.

But the work that would so profoundly influence the future of immunology actually took place after the *Princesse Alice II* returned to Marseilles in September. When the two scientists returned to Paris, they continued an active collaboration on the work begun at sea. *Physalia* being not readily available in Paris, they switched to a different organism producing a similar toxin, the sea anemone. After carrying out a number of additional pharmacological studies, the pair turned their attention to the ability of the sea anemone toxin to induce an immune response in animals. This was fairly standard scientific practice at the turn of the century, when immune responses to a wide range of bacterial toxins were being studied. It would be the key to a startling discovery. After trying unsuccessfully to immunize several different species, they began working with dogs. To their great surprise, they found that after an initial injection of toxin, rather than developing a protective immune response, the dogs became hypersensitive to subsequent doses of the toxin. The following quote from their laboratory notebook illustrates what Portier and Richet observed:

> 10 Feb [1902]—26 days after first injection—the dog was in perfect health, cheerful, active; the coat was shiny. On this day at 2 P.M. it was injected with 0.12 cc toxin per kg. Immediately produced vomiting, defecation, trembling of front legs. The dog fell on the side, lost consciousness, and in one-half hour was dead.

This finding was completely unexpected, but completely reproducible. Initially, it was interpreted as a sensitization produced by the action of the toxin itself—that is, an alteration in the animal's tissues caused by the toxin which made the animal unusually sensitive to a subsequent dose of the same toxin. Portier and Richet called this phenomenon *anaphylaxis*, from the Greek "a–" (not) and "phylaxis" (protection). A year later, however, it was shown by another French scientist that the same result could be obtained by immunizing an animal with normal, nontoxic proteins as well. As with the toxin experiments, hypersensitivity was not always induced, but it could be if the initial immunization regimen was properly managed. The experiments with normal protein antigens stunned everyone working in the new field of immunology and had to be repeated numerous times by many different researchers before they were accepted.

Clearly a new hypothesis was required. Because hypersensitivity responses to both toxic and nontoxic substances were found to be antigen-specific (that is, hypersensitivity in each case was restricted only to the immunizing antigen, and no other), it was difficult to escape the conclusion that the responses were immunological in nature. The problem was that no one wanted to believe that a physiological system designed to protect, namely the immune system, could also maim and even kill. Certainly in these early days of immunology, when no one could be quite sure what it was one was discovering, and how it would fit into the overall picture of immune protection, it was easy to set aside such unsettling observations. But in fact, these observations never did go away, and would ultimately prove to be harbingers of a darker side of the immune system barely imagined in the beginnings of this new medical science.

One of the most dramatic examples of hypersensitivity can be seen in the guinea pig. Guinea pigs are unusually susceptible to the induction of hypersensitivity. The injection of extremely small levels of things as innocuous as egg white protein can make guinea pigs hypersensitive. Subsequent injection of the same antigen will induce a state of *anaphylactic shock*. Within minutes of the second dose, the animal becomes restless, begins rubbing its nose and eyes, and experiences difficulty in breathing. Its fur stands on end; it may urinate and defecate. It hiccups violently, gasping for air. Blood pressure begins dropping rapidly, as does temperature, and the heartbeat becomes markedly irregular. Lowered blood pressure deprives the brain of oxygen, leading to disorientation and loss of muscular control. If the reaction is strong enough and the animal is not rescued with drugs, convulsions set in and death from asphyxia follows shortly. Postmortem examination shows lungs completely stretched out of shape and filled with fluid and air. Of particular importance, the openings from the windpipe to the lungs are almost completely closed off owing to contraction of the surrounding muscle. As gruesome as all these facts may seem, mark them well, for we will see shortly that the identical set of symptoms can develop in people.

So how do guinea pigs manage to survive long enough even to reproduce themselves, in the face of such violent immunological onslaughts from within? The first thing to realize, of course, is that guinea pigs in nature rarely, if ever, undergo hypersensitivity reactions of the kind generated in the lab. As some early commentators noted, the deck had been entirely stacked against the guinea pig in the laboratory by manipulating the form, timing, dose, and route of administration of antigen. Of course, they were making this point because basically they didn't like the implications of the results, but they were absolutely correct. What was so upsetting was the possibility that what was being observed in the guinea pig might also be true in humans. The believers in this possibility forged ahead with their guinea pig studies; what they learned has turned out to be critical in the understanding and management of hypersensitivity in humans.

As late as 1927, eminent authorities continued to declare that the hypersensitivity reactions observed in laboratory animals did not occur in humans. They were eminently wrong. Through careful observation in the clinic, and comparison of these observations with laboratory studies on animals, a darker truth had to be faced. The human immune system is indeed capable of turning its weapons inward against its human host. As we will see, sometimes the result is no more life-threatening than a mild (or even severe) case of allergy. The violent response to manipulated forms of antigen described earlier for guinea pigs is indeed rare in nature, for humans as well as animals. But the following two cases show that on occasion hypersensitivity reactions in humans can be quite severe.

Case 1: A twenty-five-year-old white male laboratory employee reported for work with a sore throat and was given 500,000 units of penicillin by injection into the left deltoid muscle. Although there was no history of having been treated with penicillin previously, he became apprehensive within two to three minutes, and complained of burning and tingling sensations of the scalp, "tightening" of the chest and throat, respiratory distress and headache. He began to perspire profusely, rapidly developed edema about the eyes, mouth and throat, and collapsed. Cyanosis was marked, and pulse could not be felt, nor blood pressure ascertained. A tracheotomy was performed, artificial respiration was applied, and aminophylline (0.5 mg) and epinephrine (1 ml 1:1000) were given intracardially. Oxygen was administered through a catheter. The pulse became manifest, the blood pressure could be determined, and breathing resumed. The patient remained unconscious for twelve hours. Subsequent recovery was uneventful. A history revealed that he had been working in a tissue culture laboratory handling media containing antibiotics, including penicillin, and had suffered allergic attacks during the pollen season. He was instructed to wear a dog tag thereafter inscribed with the warning that he was dangerously allergic to penicillin.

Although reported in the typically dry and detached clinical style favored by today's medical scientists, there is no doubt that the situation being described was drama of the highest order. The classical signs of anaphylactic shock were obviously spotted immediately by an alert and experienced physician who did not hesitate to take extreme action on the spot: cutting open of the windpipe to aid breathing, forced introduction of air together with pure oxygen, and direct injection into the heart of stimulants to revive and maintain an effective pulse. After a period of unconsciousness, the patient underwent an "uneventful" recovery and was sent on his way with a reminder to wear a MedAlert bracelet. The next patient was not so fortunate.

Case 2:   A thirty-year-old white farm laborer sufering from allergic asthma and hay fever was admitted to the allergy clinic. On the first occasion he was scratch-tested on the arms with a number of pollens prevalent in the area, with negative results. The next day, intracutaneous tests for sensitivity to various foods were conducted on the skin of the back. After several tests had been performed, the patient suddenly complained of difficulty in breathing and collapsed. Despite the administration of aminophylline and atropine sulfate intravenously, and epinephrine intracardially, together with artificial respiration, the patient expired within fifteen minutes. Autopsy revealed visceral congestion, edema of trachea and epiglottis, subpleural hemorrhages over the right lobe, and marked emphysema of both lungs.

The autopsy findings with this unfortunate patient are not markedly different from that for the guinea pigs described earlier, in which hypersensitivity was deliberately induced in the laboratory. The cause of death in both cases was asphyxiation due to constricted air passage to and from the lungs. Although the provoking antigen was not defined for patient number two, most likely one of the test antigens related to foodstuffs previously ingested by the patient stimulated the abrupt and violent response that led to his death. We will talk about food allergies (normally a very mild form of hypersensitivity) later in this chapter. But it is worth noting that both these patients had previous allergic disorders. This is the usual finding in cases of severe hypersensitivity problems.

Allergy in Humans: The Tip of the Iceberg

As many as one in five Americans may suffer from one form or another of *allergy*. The most common form of allergy is based on an immune reaction called *immediate hypersensitivity*. Immediate hypersensitivity reactions are called that because the symptoms manifest themselves within minutes of exposure to antigen in sensitized individuals and peak within a few hours. As we will see later, there is also a set of hypersensitivity disorders in humans called *delayed hypersensitivity* reactions, in which the symptoms may take one to two days after exposure to antigen to become apparent and may not peak for several days. As might be imagined, the underlying immune parameters are quite distinct in these two situations.

The list of substances that provoke immediate hypersensitivity responses in humans is virtually endless, and may well include almost anything in the biological or chemical environment. Of course, no one person (fortunately) ever develops immediate hypersensitivity to all possible provoking antigens (called *allergens* when we are specifically talking about antigens that induce allergic reactions). While some allergens may induce immediate hypersensitivity in large numbers of individuals—certain plant pollens, animal dander,* house dust—others are as individual as people are: specific drugs or chemicals; a particular brand of makeup; certain foods. The list of symptoms is similarly long: runny noses, itchy eyes, shortness of breath, rashes and eczema, diarrhea, and so on. Little wonder that it took many years before it was determined that all of these various problems and symptoms are related by a common mechanism, let alone that they are all caused by the body's own immune system.

As mentioned earlier, it was assumed from the beginning that immediate hypersensitivity reactions in animals are caused by antibodies. That this was also likely to be true in humans was suggested by an interesting experiment reported in 1921 by two German physicians, Carl Prausnitz and Heinz Küstner. Part of the fascination with their studies is that they carried out some of their experiments on each other. Küstner was allergic to a protein in cooked fish; Prausnitz wasn't. In addition to experiencing distressing symptoms when he ate cooked (but not raw) fish, Küstner also found that when he injected tiny amounts of a protein extract of cooked fish intradermally into his forearm, a rapid and marked reaction ensued. In about ten minutes a small welt began to arise at the site of injection. It looked very much like a mosquito bite, and it itched like one. The rapidly growing welt reached about an inch and a half in diameter and was surrounded by a red, patchy region nearly four inches across. After about twenty minutes, Küstner began to experience the more generalized, systemic manifestations of a classical hypersensitivity reaction: The itching spread to other parts of his body, he began to cough, and he had

* "Dander" is the dried, sloughed outer layers of animal skin. In animals that groom themselves using their tongue and saliva, salivary proteins may be important in provoking allergic responses in humans. Cats are probably the main source of allergenic dander for humans, and salivary proteins are definitely part of what irritates us.

difficulty breathing. After another twenty minutes the symptoms leveled off and then drifted back to normal.

The critical part of the experiment involved his colleague Prausnitz, who was not sensitive to fish in any form. When Prausnitz was injected under the skin with the cooked fish extract, absolutely nothing happened, no matter how much was injected, or how often. But if Prausnitz was first injected with a small amount of Küstner's serum, and *then* injected the following day with a bit of the fish extract, the exact pattern of local swelling and itchiness seen in Küstner developed in Prausnitz.

This is one of the most important experiments in early immunology, and it laid the groundwork for the study of allergies. It demonstrated in the clearest possible way that the agent active in causing immediate hypersensitivity in humans *circulates in the blood*. It also strengthened the connection between immediate hypersensitivity in animals and common allergy in humans. It had been shown several years before that transfer of serum from a hypersensitive animal to an unsensitized animal transferred the hypersensitive state. This phenomenon was called *passive transfer* of hypersensitivity. In animal experiments enough serum could be transferred that general (systemic) as well as local reactivity could be detected. The skin test developed by Prausnitz and Küstner provided a way to achieve the same effect in humans without having to remove acceptably large amounts of blood from a donor. Their technique was further refined as a way of routinely screening for allergy to specific substances in humans; the "P-K" test was a standard of the allergy clinic for many years.

Eventually an antibody was found that was responsible for hypersensitivity reactions. It is a special type of antibody called IgE. One of five major classes of antibodies made by humans, IgE is present in very low concentrations in the blood of normal individuals. The B cells specializing in IgE production tend not to hang out in lymph nodes and spleen, but are found in the skin, lungs, and intestinal lining—the points of entry for many pathogens. For some reason, a few individuals seem preferentially to make IgE-type antibodies in response to certain environmental antigens. The first time someone makes IgE antibodies, nothing much happens. For example, an initial bee sting may result in nothing more than the discomfort of the sting itself. But a subsequent sting from the same type of bee may result in a mild or severe hypersensitivity reaction. Why some people make IgE and some do not is not at all understood. Nor do we understand why people with the propensity to make IgE do so only in response to some antigens and not to others.

Life other antibodies, IgE by itself is relatively harmless; it simply has the property of binding tightly and specifically to a particular antigen. The reactions that lead to hypersensitivity are due to the unique homing properties of the tail portion of IgE. The initial exposure to allergen triggers the production of IgE. When the IgE antibodies build up to a critical level, they begin to bind to two special cell types, *mast cells* and *basophils*. It is these cells that actually cause the harmful effects seen in immediate hypersensitivity reactions. Both cell types are filled with granules that contain a variety of highly active pharmacological reagents, chief among which is *histamine*. Notice that the antigen binding sites on the IgE molecules point outward, away from the cell and toward its environment. They act in effect as borrowed "eyes" for the cells, helping them to survey their antigenic universe. The IgE molecules can remain sitting on the surface of these cells more or less indefinitely. Even Prausnitz and Küstner noted that Prausnitz remained sensitized by Küstner's serum many weeks after transfer. When antigen (allergen) comes into the system a second time and interacts with this surface-bound form of IgE, the basophils and mast cells are triggered to release the contents of their granules, including histamine, into the bloodstream. It is this *degranulation reaction* that leads to all the unpleasant side effects associated with immediate hypersensitivity and allergy.

We know a lot about histamine, and it is clear that together with a few other biochemical components of mast cells and basophils, histamine can account for virtually all of the sequelae of immediate hypersensitivity reactions. When histamine binds to blood capillaries, it causes them to enlarge and become more permeable to blood fluids. This is responsible for the rash (*erythema*) associated with allergic reactions that take place at the body surface. Larger-diameter capillaries simply carry more red blood cells to the affected site, causing a change in skin color from the normal pink to a deeper red. But that's just a cosmetic problem. Of much greater medical concern is the fact that the increased permeability of blood vessels, if it occurs systemically (throughout the body), will also cause a drop in blood pressure and lead to a state of potentially lethal shock.

Another problem caused by histamine is its tendency to bind to *smooth muscle*. This is a form of muscle found at many sites in the body, but most important for allergy victims, it is the muscle that surrounds the bronchioles leading into the air sacs of the lungs and that controls their diameter. Histamine causes smooth muscle to contract. In the case of bronchioles, this leads to a marked constriction, narrowing the passageway for air into and out of the lungs. Unfortunately, one of the highest concentrations of mast cells in the body is found in the lungs. When histamine is released from mast cells into the surrounding lung tissue, the resultant constriction of the bronchioles becomes a major factor in the respiratory distress, and even respiratory failure, accompanying immediate hypersensitivity reactions. Air can usually be forced in by strong, voluntary contractions of the diaphragm (gasping), but subsequent relaxation of the lungs is not strong enough to force the air back out. In the experiments described earlier on anaphylaxis in guinea pigs, autopsy usually showed distended lungs that floated in water. Asphyxiation occurs with the lungs full of used air.

Specific Forms of Human Allergy

Hay Fever. Descriptions of what is almost certainly hay fever appear in medical and nonmedical texts almost as far back as the beginning of written history. We can guess that this condition has been around for a very long time and will likely remain part of our lives for as far as we can see into the future. Serious attention to hay fever as an important and distinct clinical problem worthy of research appears to have begun after publication of a paper in 1819 by John Bostock, M.D., an English physician and physiologist. His work describes his own perennial affliction with classic hay fever symptoms. The title of his publication, *Case of a Periodical Affection of the Eyes and Chest*, shows his awareness of the seasonal association of the problem, and the particular involvement of the eyes and lungs. Unfortunately, Bostock did not understand the causative agent, believing it to be the increase in sun exposure occurring in early summer. It was not until the 1870s that the true inducers of hay fever were identified. However, Bostock's rather detailed description of his own experiences seems to have triggered a responsive chord in other medical investigators, some of whom must certainly have suffered from allergies themselves. A number of other papers followed in succeeding years, and attention soon focused on flowers, grasses, and animal danders as among the real inducers of hay fever symptoms.

Hay fever, despite its name, is not a fever and is in fact only rarely caused by hay. But it often *is* caused by pollens or other plant-associated products. Pollens are produced as part of the otherwise unremarkable mating habits of certain plants. Because pollens are frequently carried by wind from an ardent suitor to the object of its desire, allergies truly can be due to "something in the air." Like most animals, plants prefer to mate only at certain times of the year, so although allergic individuals carry their potential for allergy throughout the year, actual symptoms are usually only provoked during certain seasons. In North America, one of the most serious offenders is *ragweed*, a plant that spreads its pollen throughout much of the summer and early autumn.

Hay fever, or *allergic rhinitis*, the term doctors use, can also be caused by any airborne allergen—chemicals, dust, microbial spores, animal dander, fibers, or insect parts, in addition to pollen. As the term allergic rhinitis implies, the nose is a particularly sensitive target. The nose is unusually rich in small blood vessels and secretory glands, related to its role in warming and moistening incoming air. Even in the absence of an allergic reaction, the nose may secrete as much as a quart of water every twenty-four hours as it moistens the air passing by. Hairs in the nasal passages help trap airborne particles, and are thus a natural filter for incoming allergens. The nasal passages are lined with IgE-secreting B cells and with both mast cells and basophils, so the allergic response in the nose is both rapid and rabid. Histamine release from the mast cells causes local blood vessels to dilate and become more permeable. Fluids cross out of the blood vessels into surrounding tissue spaces, creating a sense of swelling and pressure. These fluids need to escape from the area, and in part are expressed through secretory glands and membranes, leading to an endlessly running nose. The reaction rarely remains confined to the nose, however, and usually involves the roof of the mouth and the throat (contributing to the annoying sensation of postnasal drip), and particularly the eyes, the surrounding tissues of which have their own IgE-producing B cells and mast cells.

Hay fever–like symptoms caused by other than seasonally produced plant or animal products will of course be with the poor sufferer year round. The term "hay fever" is usually applied to seasonal allergic manifestations, with the more general term *perennial allergic rhinitis* reserved for year-round upper respiratory tract allergies. Interestingly, perennial rhinitis affects females much more than males; the ratio is about three to one. The most common allergens in perennial allergic rhinitis, as mentioned earlier, are substances like animal dander or fur, airborne molds or spores, house dust (usually contaminated with dried insect parts), minute fibers from cloth, and anything else floating around in the air that is capable of calling up IgE antibodies in sensitive individuals.

Drug and Venom Anaphylaxis. The allergens associated with hay fever generally induce symptoms that are annoying but hardly life-threatening. On the other hand, a few substances can induce hypersensitivity reactions that are every bit as violent as those described earlier for laboratory guinea pigs. Among the more common allergic reactions that can result in anaphylactic shock are those to certain drugs—particularly penicillin and its derivatives—and reactions to venomous bites, particularly by insects such as bees and certain biting ants. Almost everyone has heard of someone who went into shock and nearly died as the result of a bee sting, or as the result of an unsuspected allergic reaction to penicillin.

Like hay fever, these reactions are mediated by IgE and mast cells. The reactions are swift and, if not rapidly treated, deadly. Symptoms usually begin within minutes of exposure to the allergen and may be accompanied by a range of symptoms—faintness, breathing difficulties, nausea, tingling of the skin and scalp. Extreme breathing difficulties and a drop in blood pressure are the most life-threatening symptoms and require immediate treatment. As with other forms of allergy, these symptoms do not occur on the first exposure to the allergen. The initial exposure simply builds up high levels of IgE. Subsequent exposure, particularly if the allergen is introduced into the bloodstream, provokes the anaphylactic response. In the United States, there are still several hundred deaths each year from anaphylactic shock developing in response to drugs or venoms. The formal cause of death in such cases is usually asphyxiation, or the complications of vascular collapse and shock.

Asthma. If the inhaled allergen penetrates beyond the nose-throat area into the lungs, and if there is a sufficiently high concentration of mast cells displaying allergen-specific IgE that can bind the allergen, the more serious problem of *asthma* may arise. Asthma is related to the Greek word for gasping or panting. Asthma can be caused by the very same allergens that cause hay fever. In fact, the older literature refers routinely to "hay asthma." Asthmas can also be caused by allergens that enter the body by other routes—some foods, for example, may occasionally cause asthma attacks—as long as some form of the allergen (or the IgE it induces) is able to travel through the bloodstream and reach the respiratory system. Asthma occurs in all known human populations; in its various forms it probably affects about 2 percent of the people in the United States. Although the management of asthma has improved dramatically in the past fifty years, it is still responsible for some two thousand to three thousand deaths per year, mostly among the very young and the very old.

Like hay fever, descriptions of what are clearly asthmatic attacks can be found among the oldest medical manuscripts, certainly as far back as some of the Egyptian medical tablets. Virtually all medical literature throughout the ages accurately describes at least the symptoms of this problem. Galen (ca. A.D. 130–200) thought asthma resulted from blockage of the air passages followed by fluid dripping from the brain into the nose and lungs. Maimonides wrote a long and detailed treatise on asthma in the twelfth century that would need only minor polishing to be used as a medical school text today. Although Maimonides was obviously unaware of the involvement of the immune system in allergy, he was very perceptive about many of the physiological and psychological parameters of this disease. (An immune basis for asthma was not proposed until about 1910.)

Asthma is a very complex condition that has causes other than immediate hypersensitivity, that is, other than an overactive immune response. The forms of asthma caused by inhaled allergens interacting with IgE on mast cells are referred to as *extrinsic asthma*, because they depend on interaction with substances that enter the body from the outside. But essentially the same symptoms can be caused by a variety of other factors *not* involving allergens or the immune system. Stress, for example, can be a major inducer of asthmatic attacks in certain individuals. Some of the neurotransmitters released during emotional or traumatic stress can either trigger, or certainly exacerbate, an allergic attack of any kind, but particularly asthma. Exercise is also a well-known potentiator, and possibly inducer, of asthma in sensitive individuals. Clinicians refer to these kinds of asthmatic attack as *intrinsic asthma*. In fact, many asthma attacks are a combination of the two forms, making treatment a real test of the physician's skill. Asthma is one of those diseases that highlights the delicate balance between the immune system and the brain and central nervous system, about which we shall discuss a great deal more in a later chapter.

To a considerable extent, extrinsic asthma, like allergic rhinitis, is caused by the release of histamine and other mediators from mast cells and basophils. Certainly the early stages of an asthmatic attack are closely dependent on IgE and mast-cell levels. However, other elements of the immune system are also involved in asthma, making even the immunological aspects of this disease more complex. An hour or so after an IgE-mediated asthmatic attack begins, the lungs may be invaded by white blood cells. These cells stimulate the formation of sugar-like molecules (*mucopolysaccharides*) that are secreted into the bronchioles together with the excess fluid accumulating in response to histamine.

The severe difficulty in breathing (*dyspnea*) during an asthmatic attack is thus the result of several related pathologies. Histamine causes constriction of the bronchioles, narrowing the passageways for air into and out of the lungs. The accumulation of mucous secretions caused by white cell infiltration and the buildup of fluid in the bronchioles also impedes the flow of air. Finally, histamine acting locally in the lungs leads to the accumulation of fluid in regions where the lungs normally take up oxygen from inhaled air, leading to oxygen depletion in the blood (*anoxia*).

True extrinsic (immune-based) asthma is more common in children than in adults. Up to 10 percent of preteen children may experience asthma to some degree. The vast majority of youngsters with asthma also manifest other allergies, such as hay fever, or drug hypersensitivity. Often as the child gets older, both asthma and the related allergies decrease substantially. Asthma can be a terrifying experience for both parents and children. In serious attacks, with widespread constriction of the bronchioles, it becomes very difficult for the child to expel air from the lungs, and an asthma attack can begin to approximate the anaphylactic shock syndromes described earlier in guinea pigs. Although the lungs may be full of air, not enough oxygen is getting into the bloodstream; the brain tells the lungs to try to take in more air. The result is severe gasping and wheezing. Fortunately, a wide array of highly effective bronchodilators is available at any pharmacy.

For reasons that are not entirely clear, asthma in adults is more prevalent among people of African descent. Equally mystifying is the increase—estimated to be 50 percent or more—in asthma cases generally in the past ten years.

Because of the involvement of so many components of the immune system, asthmatic attacks can last for several hours. They are usually accompanied by a range of other symptoms, differing from patient to patient: itchiness, chills, drowsiness, excessive urination. The patient is usually greatly fatigued after an attack, largely due to the heavy muscular work associated with trying to expel air. Because the symptoms are not trivial, asthma can be an expensive disease in terms of medications, doctor visits, and, in adults, time off from work. Generating some thirty million visits to the doctor annually, and several billion dollars in treatment costs, asthma is clearly a mainstay of both the medical and pharmaceutical industries in this country.

Food Allergies. What could be more central to staying alive and healthy than eating? Considering that most of us will eat twenty-five or thirty tons of food in a lifetime, the likelihood of an adverse response to at least some foods should be pretty fair. Food allergies certainly have the potential to be among the most life-threatening, or at least health-threatening, of all the immediate hypersensitivities. Although this is a possibility only rarely realized, scores of people die of anaphylactic responses to food allergens each year in the United States. Fortunately, the vast majority of food allergies lead only to nausea, vomiting, cramps, and diarrhea; they may also involve distress outside the gastrointestinal tract such as itching, hives, or asthma. This is not particularly pleasant, but it is not particularly life-threatening and is easy to avoid once the offending foodstuff is identified.

A distinction should be made between *food intolerance* and *food allergy*. The latter is a true immunological hypersensitivity to a particular food; the former includes basically everything else that causes a problem with that food. Numerous studies have shown that the majority of self-diagnosed "food allergies" are simply gastrointestinal distress identified in the patient's mind with a particular food eaten around the time the distress occurred. Usually less than one-third of these self-reported allergies holds up with controlled testing in the allergy clinic using the suspected food allergen. The prevalence of true food allergies in the general population is actually about 2 percent or less, and most of these are in children. Food allergies in adults are more rare, but as we saw in an earlier case history, they can be deadly indeed.

Almost all food allergies are to proteins. As with all food, serious digestion of protein begins in the stomach. Acid produced by stomach cells denatures (unfolds) food proteins, and enzymes in the stomach begin the process of true digestion (in the case of proteins, disassembly into its component amino acids). When the protein passes from the stomach to the small intestine, it is hit with an infusion of powerful protein-degrading enzymes from the pancreas that continue the digestive process. In a normal, healthy adult, this process will be essentialy complete—that is, the proteins taken in as food will be completely degraded into amino acids, and these amino acids will be transported across the intestines and into the bloodstream for use by the host as building blocks in the synthesis of its own proteins. Occasionally, however, very small amounts of partially digested protein may cross the intestine. In individuals with gastrointestinal disorders such as ulcers, some food proteins may even cross the gut (intestinal tract) without being digested at all. Once proteins or protein fragments large enough to be antigenic cross into the bloodstream, they are no different from any other foreign protein entering the blood and have the potential to induce an immune response.

The most common sources of food proteins causing immediate hypersensitivity reactions in humans are milk, egg (the white), peanuts, fish, and soy, more or less in that order. Allergies to peanuts and other nuts can be deadly. While the allergy is to a protein associated with peanuts, traces of this protein may be found in peanut oil, and foods cooked in peanut oil can trigger a violent allergic response. This points up one of the real difficulties in tracing food allegies. One oft-quoted case describes a violent reaction after eating a tuna sandwich. The reaction was not at all to something in the tuna; the knife used to cut the sandwich had just been used to cut a peanut butter sandwich. The poor patient nearly died!

Allergies develop more often to raw foods than to cooked foods. Food additives or preservatives may also be allergenic. As with all other allergies, allergic symptoms develop only after a second or third exposure to the offending allergen. The allergic symptoms may show up in almost any part of the body, with the digestive tract being only one of them. The underlying mechanisms in food allergy are exactly the same as in any other allergy: selective production of IgE in response to a particular food allergen entering the bloodstream, and then interaction of that IgE with mast cells. Because both the IgE and the allergen are free to travel anywhere in the body, food hypersensitivity can manifest itself in many different forms: hives, asthma, or fatigue as well as cramps, nausea, or diarrhea.

Food allergies are most common in children, particularly during the first two to three years of life. There are several reasons for this. Most have to do with the fact that at birth the human digestive system is still somewhat imperfect. There is less acid in the stomach, and fewer digestive enzymes overall. Many of the barriers to intact proteins crossing out of the intestines are not yet fully developed. Maturation of an infant's digestive system is aided by breast milk, and breast milk also brings in antibodies to help neutralize potentially antigenic substances. In most cases allergic symptoms simply disappear with time, but occasionally they will persist into adulthood. The most effective treatment for food allergy (or food intolerance, for that matter) is avoidance of the offending food. For many human infants, cow's milk can be a potent allergen; switching to either breast milk or formula usually solves the problem. The most important factor is to have the condition properly diagnosed by a doctor or an allergist. If the problem is truly a food allergy, it is a good idea to have this confirmed and recorded as part of a child's permanent medical record as it may indicate a general predisposition to allergy.

A Dream Gone Wrong: Immune Complex Diseases

We saw earlier how the discovery of antibodies by Emil von Behring at the end of the nineteenth century completely revolutionized the treatment of infectious diseases. Deadly bacteria or their toxins were injected into large animals, from whom life-saving antisera were subsequently harvested. Within a few years the mortality from diseases such as diphtheria and tetanus plummeted as antibodies produced in horses or sheep were administered to people, especially children, infected with disease-causing germs. This treatment, called *passive immunization*, was effective even for individuals in advanced stages of disease—literally on death's doorstep.

The possibility of raising huge quantities of antiserum for the price of a few bales of hay should have ended the threat of harm from infectious disease once and for all, and done away with the need for the (at the time) slightly more risky practice of *active immunization* (exposure of the individual to killed or attenuated forms of the pathogen, provoking internal production of disease-fighting antibodies—in other words, vaccination). A few of us can remember back to our childhoods and a time when antiserum treatment would commonly be used. Yet today this treatment is used only in extreme emergencies—for example, when an infection has become so overwhelming before treatment is started that the patient is in mortal danger. Why?

The problems with antiserum therapy became apparent almost

immediately. For one thing, it was initially hoped that the protection transferred with immune serum would be long term. This quickly proved not to be the case. Studies with animals showed that the protective and therapeutic effects of passively transferred immune serum wore off in a matter of days, a few weeks at best. We now know that antibodies (our own or someone else's injected into our bloodstreams) are routinely excreted through the urine. Unfortunately, an even more insidious problem awaited. Maurice Arthus reported as early as 1902 that rabbits injected repeatedly with horse serum (a common source of immune serum for use in humans) developed serious problems of skin rash, and occasionally severe local tissue damage. In some cases, repeated inoculations could induce anaphylactic shock. Because the so-called Arthus reaction was difficult to reproduce in other animals, its significance was not immediately appreciated.

However, physicians using the new serum therapy in the clinic soon began to note uncomfortable similarities between the Arthus reaction and the effects of antiserum treatment on their patients (particularly children). Therapy with specific antitoxins produced in horses and other large animals continued to reduce dramatically the mortality from bacterial infections, but the side effects were becoming too obvious and significant to ignore. The case was put clearly and forcefully in 1905 in a classic monograph summarizing experience using serum therapy to treat young patients at the Kinderklinik in Vienna. After several injections of immune horse serum for the treatment of diphtheria or scarlet fever, a fairly standard set of reactions would usually set in. These included rash, accompanied by swelling and itching, which always started at the site of injection, but which could spread to distant sites on the body. These symptoms were often accompanied by fever and swollen lymph nodes (*lymphadenopathy*), and on occasion by joint pain and reduced numbers of white blood cells. Fever, by the way, is one of the factors distinguishing this form of immediate hypersensitivity from that mediated by IgE, which results in a *drop* in body temperature.

The major contribution of the Kinderklinik monograph, however, was a forceful argument in favor of these symptoms being due to the immune response of the host to the injected horse serum proteins (antibodies) that, although providing protection against a deadly disease, were regarded as foreign and attacked by the recipient's immune system. The horse proteins, including the disease-fighting antibodies, induce the formation of counterantibodies in the human patient. Because of the large amounts of the incoming horse proteins, and repetitive exposures to them, the host antibody response builds up to a very high level and begins to cause a great deal of nonspecific damage. This response came to be known as *serum sickness*, a term that remained in use for the next fifty years or so.

The recognition of serum sickness as a frequent concomitant of passive immunization dealt a swift and final blow to the use of animal antisera for the routine treatment of human infectious disease within a decade or so of its introduction. Its demise was doubtless hastened by the death of the two-year-old child of Paul Langerhans, the German discoverer of the so-called islets of Langerhans, the structures in the pancreas that contain the cells producing insulin. The child was being treated with horse antibodies after having contracted diphtheria. After several previous injections, he was brought to the clinic for a final injection and died within minutes of receiving it. The elder Langerhans was a much loved and respected figure in Germany, and the story received extensive coverage in the press. Doctors everywhere became much more cautious about using this form of serum therapy and began once again to look to active rather than passive immunization.

Today, a combination of active immunization and antibiotics make death from most infectious diseases an exceeding rarity in industrialized countries; passive administration of antibodies is used only in extreme emergencies. Nevertheless, the drop in mortality from infectious diseases following introduction of serum therapy stands out as one of the most remarkable advances in curative medicine in human history. As we saw earlier, Emil von Behring, who died while World War I was still in progress, was eulogized by nations on both sides of this conflict as a true hero of medical science.

The basis of serum sickness is now reasonably well understood. Normally, when antigen enters the system and triggers an antibody response, the antigen-antibody complexes that subsequently form are quickly and efficiently removed from the system by macrophages and other phagocytic cells. But in those cases where the complexes build up faster than macrophages can clear them out, serious problems may develop. The buildup of these *immune complexes* on capillary walls, particularly in the kidneys and lungs, can cause the capillaries to burst and eventually lead to destruction of surrounding tissue. In the kidneys, this can lead to a condition called *glomerulonephritis*, a degenerative kidney condition that, if unchecked, can lead to loss of kidney function.

A variety of different foreign antigens may initiate immune complex disorders. Remember, it is not the type of antigen that matters, or even the type of antibody induced by the antigen. The damage comes from prolonged exposure to the antigen, and continuous induction of antibody. One of the earliest diseases to be traced to this form allergy is called *farmer's lung*. This occurs in individuals who breathe in hay dust contaminated with certain bacterial products. Once sensitized, a person breathing in contaminated hay dust will experience a rapid reaction involving the lungs that is difficult to distinguish from classical hay fever or asthma. Individuals handling pigeons have been known to develop a similar response to components in dried pigeon feces (*pigeon fancier's disease*). One class of antigens that does not induce immune complex disease is food. Even in the most food-allergic youngsters, the amount of food allergen crossing the gut into the bloodstream is well below the level required to build up excess immune complexes.

Immune complex disease is seen not only in response to foreign antigens, but also in patients with *chronic autoimmune disease*. As the name implies, autoimmune disease occurs when our immune systems attack our own molecules and tissues just as if they were foreign. In such cases, given the large volume of self materials available for the immune system to interact with, the buildup of immune complexes can be a real problem; glomerulonephritis is one of the most common complications of autoimmune disease. We will discuss this process in more detail in the next chapter.

Why Hypersensitivity?

Why do we have hypersensitivity? What possible good can it do? What is its relation to positive, protective immunity? We don't really know in every case. First of all, it is not obvious that some of the responses we have been discussing should even be thought of as "hypersensitive." Immune complex disease, for example, is not obviously an overreaction on the part of the immune system. The formation of antibody in response to antigen is what the immune system is supposed to do. There is little evidence that the immune complexes that cause this disease are formed as a result of overproduction of antibody. The problem is that antigen just keeps on coming in some situations. As a result, antibody keeps on shooting it down, and soon the sheer bulk of the debris of the battlefield overcomes the ability of the bulldozers (the macrophages) to clear it away. The consequent deposition of this debris (the immune complexes) in blood vessels and the subsequent initiation of inflammation are symptoms of the immune system being overwhelmed by antigen, not a result of overreaction or inappropriate reaction on the part of the immune system itself.

"Classical" (IgE-mediated) allergic responses are the hardest to rationalize, and for one simple reason: We do not know why IgE exists in the first place. The body has four other classes of antibody; why does it need IgE? There is reasonably good evidence that in some parasitic infections, IgE is selectively produced and may take part in clearing out the parasites. But other elements of the immune system are also called into play in these infections, and it is far from clear that IgE is critical to the host response even in cases where it is induced. Moreover, during infections with parasites, it is not just parasite-specific IgE that is elevated, but IgE in general. Thus, it is not obvious that the induction of IgE during parasitic infections is antigen-specific. In those rare individuals with a deficiency in IgE (including a complete absence of IgE), there are no detectable immunological problems with parasites or any other pathogens. Detailed studies of IgE production in vivo (in the living organism) suggest that there is a fairly sophisticated regulatory apparatus for preventing the production of IgE. That is rather bizarre; why have a class of antibody whose production the body tries to prevent? We do not do this with any other class of antibody.

Why do we even have IgE? Was there a time in our past when a much more deadly pathogen threatened us, a pathogen no longer troublesome to humans because IgE drove it completely from the scene? Is IgE nothing more than a fossil image of a dangerous episode in the evolutionary history of humans? We simply do not know. For that matter, why do we need mast cells? There are also other cells that carry out many of the functions of mast cells. One rarely if ever hears of immune deficiency diseases in which IgE or mast cells are selectively missing. Is this an indication that they are relatively unimportant in the overall scheme of immunity, such that when they are deficient or missing altogether we never even notice the diference?

Unquestionably, a good many people still die each year from IgE-mediated anaphylactic shock. Before we understood anaphylaxis, and learned how to treat it (thanks largely to work in animals), doubtless more people died. But the numbers were probably never very large, certainly not on the order of those dying at the time from diphtheria, smallpox, or the plague. And again, these are the pathogens that the immune system evolved to protect us against. Failure to respond promptly and forcefully to such pathogens means certain death for an unprotected individual and compromises the survival of the species as well. For some unfathomable reason, the immune system we ended up with has IgE as part of its repertoire. We don't know why it's there, or what good it does, but there it is. Current thinking among immunologists is that IgE-mediated allergies may just be the price we pay as a species for an immune system that otherwise does an outstanding job of keeping the species from disappearing. Nature can afford to take the long view!

FIVE

Horror Autotoxicus:
The Immunology of
Self-Destruction

Allergy and the other hypersensitivity reaction described in Chapter 4 were the first indication that the immune system, which clearly evolved to protect us from infectious disease, also has the potential to do harm. As we have seen, this was not a realization that descended lightly on immunologists. It was many years before most could accept that a system so beautifully designed to defend us could also hurt us. Ever the optimists, they assumed at first that the problem was of minor proportions. Serum sickness, being the result of human intervention, could easily be avoided or controlled, and it appeared that by judicious treatment the damage from allergy and asthma could be kept within clinically acceptable bounds. Unfortunately, as it turns out, these kinds of immunological disorders would prove to be just the tip of the iceberg. What we have come to realize in recent years is that for an increasing number of human maladies, not only is the immune system not part of the solution—it is very much the problem! It is in fact enough of a problem to have warranted the creation of a whole new biomedical subspecialty, *immunopathology*, which deals specifically with diseases caused by an eager but bumbling immune system.

In immediate hypersensitivity reactions like allergy and asthma, the damage done to the body is an accidental side effect of a vigorous attack by the immune system on something completely and obviously foreign to the body. Pollen or other allergens that come into the body could never be mistaken for part of the body itself. But in immune reactions of the type we may call *autoaggression*, the damage results from an attack of the immune system directly on self tissues. In some cases, this occurs because normal cells have been physically invaded by a pathogen and altered in some way that causes the immune system to regard the cells as foreign. But in the range of disorders we call autoimmune, for reasons we do not understand, there is no obvious extrinsic agent altering self cells. The immune system simply decides, at some point in life, that certain cells in the body are no longer self. Either way, the result is the same: an attack of the immune system, in all its destructive fury, on self.

One important way in which allergy and asthma attacks differ from the autoaggression reactions we are about to see is that the former are mediated entirely by antibody, while the latter are mediated both by antibodies and by T cells. You may have noticed that we kept referring in the last chapter to *immediate* hypersensitivity reactions. That is because there is another class of reactions that for historical reasons came to be known as *delayed hypersensitivity* reactions. These are the autoaggressive reactions caused by T cells. Because delayed hypersensitivity reactions are so important in understanding immunopathology generally, we will take just a few moments to look at how they were discovered and how they function.

Tuberculosis and DTH Reactions

Like smallpox, tuberculosis (TB) may seem to the average person today to be some sort of prehistoric monster, a disease that may once have killed millions of people but which no longer has any relevance to the world in which we live. (As we will see, however, it is a monster once again rearing its ugly head.) The origins of TB are about as obscure as those of smallpox; both became major human scourges as the need to provide labor for factories in the eighteenth and nineteenth centuries led to increasingly crowded urban areas. Both diseases are spread from person to person, so it follows that the more crowded the population, the faster these diseases will spread.

It was tuberculosis (under the then popular name *consumption*) that killed the poet John Keats in the 1820s, along with his mother and his brother, and which affected one in every ten of his fellow Englishmen. "The consumption" became a symbol of the fragility of sensitive souls (read upper classes) of the period and influenced almost every aspect of upper-class culture—a sense of fleeting mortality that found expression in art, literature, and music of the period. But TB, like smallpox, was no respecter of class, and it was certainly more devastating to the working classes living in crowded urban quarters. Unlike smallpox, which usually killed its victims quickly, TB took its own sweet time, toying with its victim for years, sometimes decades, before administering the final coup de grace. As such, it was a major cause of chronic illness, as well as death, and wreaked havoc on the economies of capitalist-oriented nineteenth-century societies. Its victims lingered on, unable to work and generate an income, but requiring care and sustenance from their families and friends.

Until the end of the ninteenth century and the formulation of the germ theory of disease, tuberculosis, like any other human malady, was viewed as a curse of God or Satan, or at the very least of unknown and uncertain origin. The first animal disease to be attributed to a specific microorganism, as we have seen, was anthrax, the natural history of which was elucidated by both Louis Pasteur and Robert Koch. But the credit for identifying the first infectious agent underlying a human disease must go to Robert Koch alone. In 1892 he announced at an international meeting of physiologists that tuberculosis was caused by a bacterium, which he named *Mycobacterium tuberculosis*. Although a handful of scientists who had taken up the new field of microbiology in the 1880s were certain that such a finding would only be a matter of time, Koch's presentation shook both the scientific and the lay worlds. The evidence he had gathered in support of his claim was irrefutable, and all those in attendance at the meeting realized they were seeing the dawn of a new era in human medicine.

The reactions we now characterize as delayed-type hypersensitivity (DTH) reactions were also discovered by Koch in the course of his investigations into *M. tuberculosis*. He showed that injection of a protein extract of these bacteria called *tuberculin* into the skin of someone who had recovered from tuberculosis caused a harmful, or hypersensitive, reaction. The skin in the area of the injection became red, itchy, swollen, and painful. The patient would also at times develop a mild fever. In animal experiments where larger amounts of tuberculin could be injected, the fever became substantial, and the skin reaction often became necrotic and ulcerous. The fever reaction suggested that substances are able to travel between the reaction site and the brain, where body temperature is regulated.

As the existence of hypersensitivity reactions became better known, the tuberculin reaction was simply lumped together with allergy, asthma, and serum sickness as another example of immune system hyperreactivity. However, the tuberculin reaction eventually proved to be quite distinct from other hypersensitivities. For one thing, all of the other reactions occurred within minutes of reexposure to antigen/allergen. The tuberculin reaction, on the other hand, could not be detected for a number of hours after injection of antigen, peaked thirty-six to forty-eight hours later, and might not subside completely for several days. The delayed nature of this reaction was generally recognized, but it was not at first considered sufficient reason to view the tuberculin response as fundamentally different from other hypersensitivity reactions. In addition to the tuberculin reaction, DTH came to be recognized as the basis for a number of skin allergic reactions, like poison ivy, oak, or sumac.

A major surprise came when it was learned that, unlike all other hypersensitivities, the delayed response to tuberculin could not be transferred with serum, either in animals or in humans. When, for example, serum from a tubercular guinea pig was transferred to a healthy animal, the recipient showed no skin reactivity at all when tuberculin was injected intradermally. The inability to transfer delayed hypersensitivity with serum distinguished DTH reactions as a truly separate immunopathology. It also created a truly major dilemma. By 1930 or so, most scientists believed that, given the exquisite antigen specificity of immediate hypersensitivity reactions, these reactions must be immune in nature and thus mediated by antibody, even though no specific form of antibody responsible for allergy or serum sickness had yet been identified. As antibodies are found in serum, the inability to transfer delayed hypersensitivity with serum essentially ruled out antibody as a causative agent. Yet DTH reactions, like immediate hypersensitivity reactions, were shown to be absolutely antigen specific. If antibodies were the only agents known to have the property of specific antigen recognition, how then could DTH be antigen specific?

This dilemma was resolved by a milestone experiment carried out by Merrill Chase at the Rockefeller University in the 1940s. He showed that antigen-specific delayed hypersensitivity could be transferred between animals using cells (lymphocytes) from the hypersensitive animal. This experiment is one of the most important in immunology because it established for the first time that cells, as well as antibody, can have the property of antigen recognition. This experiment provided the foundation for the beginning of a major subdivision of immunology called *cellular immunology*. Cellular immune responses would eventually be recognized as the basis for a wide range of immunological phenomena, including (in addition to DTH reactions) transplant rejection, suppression of viral infections, many autoimmune diseases, and some aspects of tumor control. All of these reactions are now known to be caused by T cells.

Tuberculosis as an Autoaggressive Disease

As with the IgE-mediated hypersensitivities, delayed hypersensitivity can be quite deadly. Nowhere is this more evident than in tuberculosis itself. The bacteria that cause TB (M. *tuberculosis*) are spread from one individual to another in aerosol form. One individual releases them by coughing or sneezing. Being very light, the bacteria may stay aloft in the air for some time before settling out on various surfaces, where they will probably die. But while still airborne, they may be breathed in from the surrounding air by a complete stranger, and settle into the lungs. The new host is of course not defenseless; the *tubercle bacilli*, as they are called, are immediately engulfed by macrophages that live in the lung, rather than in lymph nodes. Just as with macrophages in lymph nodes, the lung macrophages' principal task is to eat everything in site. Depending on the infectious strength (*virulence*) of the invading bacilli, and the defensive strength of the macrophage, the bacilli will either be destroyed by the macrophages' digestive system or they will somehow survive and begin to replicate *within the macrophages*. This ability of certain bacteria not only to avoid being digested by macrophages but to actually take up residence within them and use them as a source of food is one of the most insidious and dangerous tricks developed by pathogenic microbes. It is a bit like having a night stalker take up residence in your attic, crawling around through your walls and ceilings while you sleep, raiding your pantry and larder while waiting to do you in.

If the bacilli do take up residence inside the macrophages, they will continue to grow until, at some point, the macrophages burst and release hundreds of fresh bacilli into the surrounding tissue. This attracts even more macrophages to the site, which promptly ingest the newly released bacilli. Strange as it may seem, the disease can actually stabilize at this point, if the new macrophages manage to kill off most of the bacilli released from other macrophages. In some cases, however, the high local concentration of macrophages attracts the attention of T cells passing through nearby blood capillaries. The T cells cross out of the bloodstream in response to chemical signals from the macrophages, and follow these signals to the source. The job of the T cells is to examine closely macrophage surfaces for the presence of nonself material.

If they find foreign matter there, the T cells release a collection of soluble proteins called *cytokines*. Cytokines do several important things. First, they attract yet more macrophages to the infected part of the lung. Second, they encourage macrophages already at the site to stay there, and not wander off. Finally, and perhaps most important, they stimulate all macrophages at the site to a veritable feeding frenzy, increasing their ability to ingest and destroy tubercle bacilli perhaps a hundredfold. The "activated macrophages" begin to lay down scar tissue around the infection, trying to wall it off from the rest of the body. They even send chemical signals to the brain, asking for an increase in body temperature (fever), to help fight the bacteria. All of these activities comprise the early part of DTH. In many cases, this may be sufficient to arrest progress of the disease and completely clear the bacilli from the host.

At this point, the DTH response could well be considered protective and beneficial. The situation turns ugly only when this early, more measured response fails to rid the body completely of all traces of the bacilli. When the T cells sense that, in spite of their efforts to attract and stimulate macrophages, the infection is persisting, they begin killing the infected macrophages to deprive the bacilli of a place to replicate. This has the unfortunate consequence of releasing the bacilli into the surrounding lung tissue, where they continue to replicate and, unhindered now by hungry macrophages, begin to infect healthy lung tissue. The T cells then proceed blindly to kill off those infected lung cells not already destroyed by the bacilli replicating inside them. There follows the disease stage with the ominous name "liquefaction and cavitation." Large sections of lung tissue are literally melted away by disease. But notice that it is not necessarily the bacteria themselves that are the major culprit; *the vast majority of the damage is done by the host's own T cells*, in what is essentially a prolonged, chronic DTH reaction. This is by no means a phenomenon restricted to tuberculosis; as we shall see shortly, it is distressingly common.

As a footnote to this story, we should note that after twenty or so years of decline in both incidence and mortality, tuberculosis is once again on the rise around the world. About three million people worldwide will die this year from tuberculosis. In the United States, we may see twenty-five thousand new cases, with perhaps two thousand or so dying from the disease. Part of the explanation is doubtless the appearance of AIDS (Chapter 6), which destroys the T-cell system and renders individuals more susceptible to diseases like tuberculosis. Other factors may also be involved. A recent analysis by the Centers for Disease Control suggests that one of the major reasons for increased mortality in recent years (as opposed to increased incidence) has been non-compliance with physician-recommended treatment for tuberculosis. At least some people stricken with this disease are just not taking it seriously. This alarming trend will call for renewed efforts by public health and health care delivery professionals to prevent its continuance; moreover, the public's understanding and support are needed if tuberculosis is not to become once again the dreaded killer it was at the end of the nineteenth century.

Viral Hepatitis

In the case of tuberculosis, the invading tubercle bacilli certainly cannot be considered harmless. If left unchecked by the macrophages, they would doubtless destroy lung tissue on their own. One could argue that if the pathogen is going to kill someone anyway, not much is lost if the immune system kills the person while trying to clear the infection. In that light, the subsequent overreaction by the T cells is perhaps understandable and forgivable.

But in the case of infection of the liver with the hepatitis B virus (HBV), it is a little harder to be so understanding. HBV-induced viral hepatitis (also known as serum hepatitis) is truly the modern equivalent of smallpox. It affects more than 300 million people worldwide, and is today one of the world's leading causes of death from infectious disease. It spreads from person to person mostly via contact with body fluids such as saliva, blood, vaginal secretions, or semen. Like HIV (human immunodeficiency virus), HBV spreads rapidly among male homosexuals and intravenous drug users, 80 percent or more of whom show evidence of exposure to the virus (compared to 5 percent in the rest of the population). It induces both an acute and a chronic form of hepatitis, either of which can be fatal, and it is also a leading cause of liver cancer. The initial symptoms are usually quite mild, barely more than a mild influenza. It is nonetheless virtually impossible to treat. The course it takes is entirely dependent on how the immune system decides to deal with it.

Although only one of several viruses that can cause liver disease, HBV is by far the most damaging to human beings. The liver damage in HBV hepatitis can be massive and devastating. Yet so far as we know, *HBV itself causes no harm at all to liver cells*. Outside the body, liver cells infected by HBV get along just fine; there is no sign of virus-induced damage. All evidence suggests that in this disease, when serious damage occurs, it is *solely* the immune system (and in particular the T cells) that causes the damage.

In tuberculosis, remember, the pathogenic tubercle bacilli invaded macrophages; in the course of trying to destroy the pathogen-altered macrophages, the T cells ended up destroying the lungs. A similar but even more deadly sequence of events takes place in HBV-induced hepatitis. Like most viruses, HBV invades normal cells and takes over the cell's machinery so as to make more HBV. In the course of doing so, the HBV integrates its own small piece of DNA into the infected cell's DNA. Once this happens, the cell treats the viral DNA just like its own. It copies out the HBV instructions for making more HBV, and at the same time copies out viral instructions that shut down many of the cell's own normal functions. This is much sneakier than tubercle bacilli, which we previously likened to a prowler crawling around in the attic and raiding the pantry. Invasion with a virus like HBV is much more like someone living inside your own skin, taking over your body and pretending to be you while trying to kill you: a very clever, and potentially very deadly, strategy.

But the vertebrate immune system has developed a rather clever strategy of its own for dealing with viral invaders. Macrophages, as we have just seen, eat everything around themselves, and display fragments of what they just ate on their surface. This tells T cells what has come through the blood and lymph lately, like bacteria. But in addition to that, every single cell in the body is required to display on its surface samples of the proteins they are currently making. This is how T cells spot viruses that have wormed their way inside a cell. The virus will make its own proteins needed for its own reproduction. The cell will send these out to the surface, just like any other protein the cell is making. The protein samples are examined periodically by patrolling T cells. As long as the proteins are samples of self, the T cells go on their way. But if the samples are not recognized by the T cells as self, then killer T cells will attack the cell and destroy it. The viruses are thus deprived of a place to replicate. If they spill out and infect a neighboring cell, that cell too will be killed. The problem is, the immune system has absolutely no way of knowing whether the virus invading the cell is harmful or not; T cells have been selected over evolutionary time to simply destroy *any* cell inhabited by anything not self. As we said before, the immune system is basically blind; it is incapable of making decisions, and so simply errs on the side of caution. This cautious approach may cost you your life!

In HBV-induced hepatitis, most HBV-infected cells meet precisely the fate just described. In the acute form of the disease, the response by T cells is vigorous, and the infection is often completely cleared. The resultant immune damage to the patient can be quite severe, but it is repairable and only rarely fatal. But in a certain number of cases, the disease is not resolved at the acute stage and progresses on into the chronic form of HBV hepatitis. This is where the greatest damage is done. The viral DNA continues to direct production of low levels of viral proteins, which make their way to the surface of infected liver cells. And the T cells just keep on killing infected liver cells. Eventually this can lead to a state called *cirrhosis*, which is a general term referring to massive liver cell destruction. It is a bit like the "liquefaction and cavitation" reaction seen in tuberculosis, and is caused by the same thing: relentless destruction by T cells.

The damage in viral hepatitis is also similar in outcome to that seen in alcohol- or drug-induced cirrhosis of the liver. Because the liver (uniquely among tissues in the body) has a certain capacity for self-regeneration, the damaged liver constantly tries to replace damaged cells with new ones. But these too become infected as HBV spreads slowly throughout the liver, creating an ongoing cycle of destruction and renewal. Unfortunately, over time the renewed liver tissues become more and more abnormal, failing to carry out their routine functions such as metabolism of food and the production of blood coagulation products and bile. In some cases destruction simply outpaces renewal, leading rapidly to liver failure and death. In other cases the constantly replicating liver cells become cancerous; they start to grow rapidly and without control. In a high percentage of advanced cases, particularly in third world countries where the necessary intensive care is unavailable or inadequate, the result is death.

Immunological Tolerance

The very first autoaggressive diseases to be detected, and understood for what they are (the immune system attacking self), were those of the type just described. The idea that a pathogen might alter "normal" self in such a way as to make it seem foreign to the immune system did not require a great deal of rationalization. And in fact, prior to about 1950, it was assumed that *all* autoaggressive immune reactions resulting in harm to human beings occurred as a result of the immune system attacking self cells altered or damaged by invading pathogens. In terms of the actual damage done, it is very difficult to distinguish the harm done to self in the course of an overreaction to a pathogen from that seen in true autoimmune disease.

In the early days of immunology, before the rules of the game were completely understood, the possibility that the immune system might also produce antibodies to self components seemed still open. At just about the turn of the twentieth century, Paul Ehrlich, one of the truly great thinkers in immunology, was amusing himself by injecting red blood cells from one goat—the "donor"—into other goats from his institute's herd—the "recipients." After a few weeks, he was always able to find in the recipients' serum antibodies that could cause fresh samples of the donor's red cells to clump and then disintegrate. But whenever Ehrlich injected donor red cells back into the donor goat, he could never find any such antibodies made against the red cells, no matter how long he waited. This seemed very curious indeed to Ehrlich. Why would the very same red cells provoke antibody formation in all the other goats, but not in the donor? The existence of such "autoantibodies" did not seem a priori forbidden to Ehrlich; in fact, given their properties, they could be quite useful. They might, for example, play some role in controlling the total number of red cells present in the body, or perhaps in the disposal of worn out and useless red cells. Yet try as he might, he could never demonstrate their existence, and he finally gave up.

Ehrlich incorporated his findings with the goat red blood cells into a concept he called *horror autotoxicus:* The body has a natural aversion, he postulated, against producing antibodies recognizing self components, because these might be toxic to self. The danger of toxicity must outweigh any advantages that such antibodies might otherwise have. When hypersensitivity reactions were discovered, he felt they simply proved his point; immunity to self would have only dire consequences. The concept of *horror autotoxicus* became one of the founding principles of immunology, assuming its place as one of the intellectual yardsticks against which all new immunological thought was measured. The possibility that the immune system could react against self was for many years simply assumed to be ruled out.

But how does the immune system learn what is self and what is foreign? How does it avoid making antibodies to self? Is this something programmed into our genes, or is it a more plastic property that can change to meet constantly evolving pathogenic challenges?

One of the earliest observations shedding light on these questions was made by Ray Owen in 1945. Owen was studying a certain type of twin in cattle called a *freemartin*. Freemartins are very interesting. They are genetically distinct twins ("fraternal" twins) that, unlike usual fraternal twins, share a common placenta. The fact that they are connected by blood during fetal life means that in utero they share everything that moves around in the bloodstream. For example, if one is male and one is female, it is possible to examine the possible influence of things like hormones produced by the male twin during fetal life on the subsequent development of the female twin. It is also the case that the cells of the blood of the two twins mix freely prior to birth, and at birth each twin has a mixture of two genetically different types of blood. Because the blood *stem cells* also mix between the twins, this state of mixed blood types persists for life.

This was a classic case of an extremely important point staring one right in the face; look a little to either side, and you would miss it. But Ray Owen didn't. He thought about his freemartins and realized that if each had its own placenta, and their bloods had *not* mixed before birth, then as adults they would surely be intolerant of each other's blood. This is what we see in human fraternal twins: Unless they happen by chance to have the same blood type, it is no more possible to exchange blood between two fraternal twins than it is to exchange blood between any two randomly selected individuals, unless they accidentally have the same blood type. But freemartin twins are completely tolerant of each other's blood, no matter how genetically disparate, all their lives. Thus emerged what would become one of the most important theoretical principles of immunology. Anything we are exposed to *prior to* birth will be regarded as self; if we are exposed to the same things *after birth*, they will be foreign.

This principle has been shown to be true in a great many laboratory experiments since Owen first reported his observations with freemartins. In mice and rats, as it turns out, the period during which prenatal tolerance can be induced actually lasts until one or two days after birth, making such experiments relatively easy to perform. For example, a newborn mouse injected with cells from an adult rat can, as an adult mouse, accept a skin graft from the same type of rat with no sign whatever of rejection. The rat skin and the accompanying fur, even if of a different color than the mouse's own, will last for life. The same piece of rat skin placed on an untreated mouse would be rejected almost immediately.

Experiments like this do not really have any practical application, although as we shall see a bit later it was the result of just this experiment that prompted the first successful kidney transplant in humans. But such experiments do reveal an important point about the process of self-tolerance. Major decisions about what is self and what is not are made while the fetus is developing inside the womb. At or near the time of birth, the newborn (or the almost-born, depending on the species) animal takes one last look around, and basically says: "Okay, this is it; this is me. Anything other than this that I see from now on is foreign, potentially harmful, and must be eliminated." This decision is communicated to the animal's immune system, which imprints it onto the T and B cells that are charged with making the self–nonself determinations. And since, as we have seen, T and B cells live only a few weeks at best before they die and are replaced, each new generation of T and B cells produced by this animal for the rest of its life will have to be imprinted with the same information over and over again, without making a single error. If this process is perturbed in any way, the result may be autoimmune disease—not an accidental spillover of damage in the course of trying to remove a cryptic pathogen but true aggression against perfectly normal, healthy self cells. In the past several decades, scientists and physicians have learned that this happens more often than we might like to think.

Autoimmune Disease in Humans

A number of human diseases that would eventually be recognized as autoimmune—that is, where the antigen that is recognized by the immune system is truly self, and not of pathogenic origin—were described in the early part of this century without a full realization of their autoimmune nature. For example, in the 1930s it was found that monkeys immunized with ground-up brain cells from other monkeys would develop what appeared to be an immune response to their own nerve tissues; the damage mimicked exactly that seen in naturally developing encephalitis. This should have stimulated a serious consideration of an autoimmune basis for encephalitis. Unfortunately, no antibodies to brain tissue could be found in these monkeys. The involvement of T cells in DTH immune reactions was not yet suspected in the 1930s. Thus, this forerunner of a classic T-cell–mediated autoimmune reaction took its place on the shelf with other unexplained immunological oddities of the day.

Although the possibility of true immune reactions against unaltered self continued to be discussed into the early 1950s, it was with the publication of a paper by Witebsky and Rose in 1956 that the field of autoimmunity was finally established. Noel Rose, a medical student at the time, was working on a research project in the general area of immunology. Taking the approach used earlier to induce encephalitis in monkeys, young Rose immunized rabbits with a preparation of purified rabbit thyroglobulin. Thyroglobulin is an iodinated protein related to the thyroid hormone *thyroxin*. It is stored in the thyroid gland and used to produce thyroxin when needed. Under normal conditions, the body certainly does not regard thyroglobulin as foreign. But when Rose removed thyroglobulin from thyroid glands, purified it, and injected it back into rabbits, the result was little short of astonishing: the development of classic thyroiditis, with exactly the type of damage to normal healthy thyroid glands seen in cases of human thyroiditis. Moreover, the thyroiditis in this case was accompanied by the production of antibodies that could be shown to be specific for normal, healthy thyroglobulin. Witebsky and Rose did not have to wait for the discovery of T cells to explain to themselves what they had observed. Rose went on to propose in a clear and forceful way that true autoimmunity did in fact exist and should be explored seriously as a potential human medical problem. He himself dedicated the rest of his career to the study and treatment of human autoimmune diseases.

The human clinical equivalent of the autoimmune thyroiditis produced in rabbits is *Hashimoto's thyroiditis*. First described in 1912, Hashimoto's disease appears most often in women over forty, causing *goiter* (enlargement of the thyroid gland in the neck), and general thyroid insufficiency. The cause was for many years unknown, although the thyroid, whenever examined at autopsy, was usually found to be infiltrated by lymphocytes, suggesting a possible immune basis. Shortly after Witebsky and Rose published their animal studies, it was shown that patients with Hashimoto's thyroiditis have antibodies in their circulation that react with thyroglobulin. This was the very first time that both a specific immune mechanism (antibody) and a specific self antigen (thyroglobulin) could be shown to explain fully a human disease (thyroiditis). For some reason, a protein (thyroglobulin) that is normally regarded as self suddenly looks foreign to the immune system. How does this happen?

Autoimmune thyroiditis would by no means prove to be the only human disease with an autoimmune etiology. The list of such diseases has grown quite long (Table 5.1), and may still be growing. Nevertheless, as we have seen, it would be many years before the notion of truly autoimmune diseases was completely accepted. The medical literature of the 1960s and 1970s is filled with references to so-called autoimmune diseases. Most scientists could not shake the suspicion that most or all of these diseases would eventually be attributable to some cryptic pathogen in the affected tissue.

Table 5.1 makes clear that almost every organ and tissue in the body can be a target for autoimmune disease. And while Table 5.1 makes a distinction between relatively organ-specific autoim-

TABLE 5.1. A Partial List of Human Autoimmune Diseases

Disease	Target organ
Autoimmune diseases affecting a relatively restricted range of tissues	
Diabetes (insulin-dependent)	Pancreas
Autoimmune (Hashimoto's) thyroiditis	Thyroid
Autoimmune hemolytic anemia	Red cells
Myasthenia gravis	Thymus
Multiple sclerosis	Nerves
Addison's disease	Adrenal glands
Crohn's disease	Bowel
Autoimmune hepatitis	Liver
Autoimmune nephritis	Kidney
Pemphigus	Skin
Graves' disease	Thyroid
Autoimmune diseases affecting a wide range of tissues	
Systemic lupus erythematosus	
Rheumatoid arthritis	
Scleroderma	
Sjogren's syndrome	

mune disease, and systemic disease that involves multiple organ
systems, in fact almost all autoimmune diseases affect more than
one system in the body. The "relatively restricted" diseases are
just that—*relatively* restricted. Patients with insulin-dependent
(Type 1) diabetes, for example, almost always have other autoimmune problems. The spectrum of diabetes-associated autoimmune diseases (pernicious anemia, Graves' disease, Hashimoto's
thyroiditis, to name just a few) is so broad that sometimes it is
easier to think of diabetes as just one part of a broad-spectrum
"pan-autoimmunity" that happens in a particular individual to
affect the pancreas more than other organs.

Those diseases that clearly affect many different tissues in the
body, such as lupus (systemic lupus erythematosus), Sjogren's
syndrome, or rheumatoid arthritis, have one peculiar feature in
common: They tend to affect women much more than men.
Whereas diabetes affects men and women more or less equally,
arthritis is two or three times more frequent in females; lupus six

to ten times more. Even some of the relatively restricted autoimmune diseases, like myasthenia gravis (which we will talk about shortly), affect predominantly women.

In those diseases affecting women most strongly, there is a tendency for the disease to appear relatively early in life, particularly during the childbearing years. There has been speculation that women are more prone than men to develop autoimmune disease because they have developed more powerful immune systems in connection with their childbearing function. Whatever the reason, it is clear that such autoimmune diseases are regulated by sex hormones. Studies in strains of mice in which the females spontaneously develop a lupus-like disease have shown that castrating males of the same strain, or injecting them with female hormones, leads to a rate of disease equal to that of females. Similarly, injecting the female with male hormones prevents, or at least limits the severity of, the disease.

In addition to the gender bias, most autoimmune disorders appear to have a genetic basis, in that they tend to "run in families." But the genetic link is only partial. In studies of genetically identical twins, for example, only a third would both have multiple sclerosis; half might have diabetes; and a quarter would both develop systemic lupus erythematosus (SLE). At autopsy—in the case of accidental death, for example—the apparently healthy twin may show subclinical signs of the disease, but it is clear that factors in the environment are very much at play in the development of full-blown autoimmune disease. And finally, as we will discuss later, there is very definitely an interplay between the mind and the immune system in autoimmunity. Hence, these are very complicated conditions indeed. Just talk to the 5 percent or so of Americans who suffer from them!

To get a feeling for the range of disorders with an autoimmune basis, let us take just a brief tour of a few of the major human autoimmune diseases.

Autoimmune Hepatitis. Earlier in this chapter we saw the immune damage to the liver that can result from infection with the apparently harmless hepatitis B virus (HBV). Interestingly, almost exactly the same disease can develop in the complete absence of any trace of virus, and it is now accepted as being a true autoimmune disease. Autoimmune hepatitis occurs about eight times more frequently in women than in men and is found almost exclusively in women of north European descent. The symptoms are essentially the same as in viral hepatitis: fatigue, weakness, jaundice, dark urine. In addition, young women with this disease usually have disturbances with their menstrual cycles. The disease results when for some unknown reason the immune system begins to regard certain proteins found on the surface of liver cells as foreign, and T cells begin to attack and destroy the liver cells. Antibodies are also formed to liver cells, as well as to muscle and even kidney tissue. If not treated properly, autoimmune hepatitis can progress into exactly the same kind of cirrhosis seen in viral hepatitis and can easily be fatal.

Several things do distinguish the autoimmune form of hepatitis from HBV hepatitis. First, even the most sensitive tests fail to detect any trace of virus. Second, it is almost always accompanied by other autoimmune symptoms such as thyroiditis, arthritis, or myasthenia gravis. And third, autoimmune hepatitis responds very well to corticosteroids, whereas this drug has minimal impact on viral hepatitis. But these differences are fairly subtle, and it takes an alert and well-trained physician to make the proper diagnosis. It was many years before an autoimmune form of hepatitis, developing in the complete absence of any extrinsic pathogen, was recognized and accepted for what it is.

This is a perfect example of why it was so difficult for both scientists and physicians to believe that autoimmune diseases are really, truly autoimmune, and not an attack on cells harboring faint traces of some hard-to-find virus or bacterium. Even today, some textbooks still hedge and hint at the possibility that autoimmune hepatitis could be due to an undetectable pathogen. But in fact, scientists have now isolated the provoking antigen in autoimmune hepatitis; it is called "liver-specific protein," or LSP, and is a perfectly normal part of healthy liver cells. And as we have seen, thyroiditis can be induced by immunization with pure thyroglobulin. Clearly, in cases such as these, the resulting autoimmune condition was not caused by an infectious agent.

Systemic Lupus Erythematosus. Nearly everyone knows of someone with SLE, or "lupus." It is the most prominent example of an autoimmune disease in which the immune system attacks not a specific tissue or organ in the body, but rather a wide range of self tissues. Like most autoimmune diseases of this type, SLE is seen more frequently in females, most often setting in between the teen years and middle life—that is, during the peak reproductive years. The *erythematosus* in SLE refers to a rash that often breaks out on the face, particularly around the nose. This so-called butterfly rash is very sensitive to ultraviolet light, including sunlight. It is, however, simply the most visible manifestation of eruptions that break out on the surface membranes of organs throughout the body. Other symptoms include fever, weakness, anemia, and kidney problems. Joint pain from arthritis is a common concomitant of lupus throughout all its stages.

Lupus is accompanied by antibodies to a wide range of self antigens, one of the most unusual being DNA (deoxyribonucleic acid), which is the genetic blueprint stored in chemical form in each cell in the body. It directs every activity of the body, including the immune system. Although many other autoantibodies (e.g., against thyroid or liver tissue; muscle; the blood cells) are found in lupus patients, antibodies to DNA are the most prominent, and are in fact diagnostic for the disease. Because DNA is buried deep inside the cell, in the nucleus, it is not easily reached by antibodies. It is likely that the DNA antibodies are formed against DNA released by dying cells. It is not clear whether the DNA antibodies themselves cause any harm. Such antibodies are not formed in other diseases in which cells die and release their contents, so their appearance in lupus is still something of a mystery. One very much has the feeling that if we knew why these particular antibodies were formed in the first place, we would understand a great deal more than we do about this disease.

Like other antibody-mediated autoimmune disorders, much of the serious damage in SLE comes from the deposition of antigen-antibody complexes in blood vessels throughout the body. When this occurs in blood vessels in the kidneys, for example, a condition known as *glomerulonephritis* can develop, which eventually may lead to serious kidney problems and even kidney failure. Because the antigens in lupus (and other autoimmune diseases) are a part of self, there is in effect an endless supply of them, and an endless stream of immune complexes just keep on forming. In advanced cases, lupus may also affect the nervous system. This can result in pain throughout the body, but may also result in actual damage to the central nervous system, manifesting as headache, paralysis, seizures, or other neuropsychiatric problems.

Like most generalized autoimmune conditions, lupus is not really curable. It can be controlled in many cases with steroids such as prednisone, but these are not without their own risks. Arthritis and kidney problems often worsen with age, causing considerable distress and affecting the general quality of life. On the other hand, it is not obvious that life span per se is greatly affected by diseases such as lupus.

Myasthenia Gravis. Myasthenia gravis is a disease characterized by extreme muscular weakness, usually beginning in the head and neck but in most cases extending to the entire body. It is twice as frequent in women as in men, and is seen earlier in women (average age of onset twenty-eight years in women vs. forty-two years in men). The disease in men is often more limited as well. The first visible signs of myasthenia are usually drooping eyelids and sagging neck and facial muscles. Patients may experience difficulty in breathing and swallowing, and may have vision problems as well.

Myasthenia was recognized as far back as the mid-seventeenth century as a distinct condition, although its autoimmune basis could not of course have been known. The following description, written by the English physician Thomas Willis in 1672 in his *De*

Anima Brutorum, points up an affliction that often accompanies the onset of this disease: "[s]he for some time can speak freely and readily enough, but after she has spoke long, or hastily, or eagerly, she is not able to speak a word, but becomes mute as a fish, nor can she recover the use of her voice under an hour or two."

This describes almost perfectly a modern-day student named Abby who enrolled in an immunology course I taught as an assistant professor at UCLA. It had not occurred to me that students would seek out this course in order to better understand their own afflictions, but that would turn out to be a fairly common occurrence over the years. Although I knew relatively little about myasthenia gravis at the time, I would learn a great deal over the following year, and it has remained a lifelong interest. Abby was a lovely young woman whose overall appearance was clearly marked by her condition. The muscles of her face had weakened considerably, although she was still very pretty. She had a heavy surgical scar peaking up over the collar of her blouse, the origin of which we shall discover shortly. Abby showed up at my office hours almost every week, asking perceptive questions about a wide range of immunological issues. She was eager to learn as much general immunology as she could so as to prepare for a research career in mysathenia gravis. At her urging I helped her get into the primary scientific literature on the subject and helped her interpret some of the topics that were beyond what we had covered in the course lectures. Abby would often become excited during some of these conversations, and particularly if it were late in the day, she could very suddenly become exhausted and unable to speak further. This was alarming at first, but she assured me that if she could just rest a bit then everything would be fine, which it always was. She finished the course with an A and went on to graduate school in immunology somewhere on the East Coast.

Prior to the mid-1930s, Abby's outlook for a reasonably normal life would have been much dimmer. It was only in 1934 that drugs that relieve the most severely debilitating symptoms of myasthenia gravis (MG) were discovered. With the development of artificial respirators a few years later, the world saw a rapid drop in mortality from this disease by 1940. Before that time patients went largely untreated and often died from respiratory failure within a year or so of onset. Currently, MG is fatal in only about 10 percent of those afflicted, although it is never curable.

The defect in MG is an interesting one; it involves one of the most highly restricted anti-self attacks of any of the autoimmune diseases. Patients with MG make antibodies that affect the response to a neurotransmitter called *acetylcholine* (ACh), which is released from the tip of a nerve cell at the point where it attaches to a muscle and is picked up by a special acetylcholine receptor (AChR) on the muscle being served. This causes the muscle to contract and carry out its function. MG patients make antibodies to the AChR; these antibodies block the muscle's ability to pick up and respond to ACh. There is nothing wrong with the muscle per se; it simply cannot be stimulated by the nervous system to do its job. In animal models of this disease, passing the antibody from an animal with MG to a healthy animal is sufficient to pass the disease. A pregnant woman may pass the antibodies to her developing child, which may be born with symptoms of the disease (the symptoms fade within the first year of life). Thus, in this instance a single antibody, specific for a single target molecule (AChR), appears sufficient to explain an entire disease.

Yet, despite the narrowness of the immune attack in MG, most patients will show some sign of a more generalized autoimmunity. As many as a third will have clinically detectable Graves' disease, which affects the thyroid. There is little to suggest Graves' disease is caused by the same antibodies that cause MG; if it were, then *all* MG patients should have Graves' disease.

There is a peculiar structural abnormality of the thymus that usually accompanies MG. The thymus, remember, is the organ that produces T cells. Ordinarily it does not have any of the cells that make antibody (B cells). But in MG patients, for unknown reasons, the thymus enlarges and acquires significant numbers of B cells. In fact, in some ways the thymus begins to look a bit like a lymph node, with highly organized regions of B-cell activity. It has been found that removal of the thymus in MG patients usually provides a marked degree of relief from some of the more severe symptoms of this disease. That was the origin of Abby's scar mentioned earlier. The relation of this feature of MG to the antibody attack on AChR is a complete mystery; in fact, it makes no sense at all. Like many other organs, there are at least low levels of AChR present in the thymus, but why would the disease either start or end in the thymus? Like the DNA antibodies in lupus, the thymic abnormalities in MG are telling us there is something we still do not know about myasthenia gravis.

Why Are We so Self-Destructive?

Why do we have these problems? Why does the immune system do this to us? How could we spend millions of years of evolutionary time and energy and come up with a system that does us so much harm?

Part of the dilemma for the immune system may well have its origin in our success as a species in other areas. Barely a hundred years ago, the immune system had to work time-and-a-half just to keep us alive long enough to find a mate. Today, most of the diseases the immune system evolved to protect us against can be controlled to a large extent by other means, such as hygiene, public health measures, or antibiotics. Hypersensitivities, allergies, damaging overreactions to harmless microbes, and autoimmunity may today seem like serious medical problems. For 99.99 percent of human existence, they went virtually unrecognized, simply because in the context of rampant infectious diseases that routinely decimated entire populations, they were scarcely discernible. Given what the immune system has had to overcome to get us to this stage in our evolutionary history, the problems that we now call immunopathologies can hardly be used to label the immune system a failure.

So one result of the success of our immune systems (together with other factors relating to nutrition, as well as relative isolation from predators and other elements of our natural environment) is a greatly extended average life span. In the past hundred years, although the maximum human life span still appears to be fixed at 120 years or so, the *average* life span has nearly doubled, owing largely to a reduction in mortality from childhood infectious diseases. Most animals in the wild live only a short time beyond the peak breeding years for their species. It may not be very flattering to our egos, but nature does not really have a role for any of us beyond the passing on of genes. We in fact become a potential problem for the next generations of breeders and their offspring by consuming valuable resources needed by younger members of the species for reproduction. The immune system, like other life-support systems, is designed to protect us up through the end of our active breeding and child-rearing season in life. It has not the foggiest idea what to do with us beyond that. True, some autoimmune problems like lupus and diabetes can affect the young, but by and large, many of the problems caused by the immune system as we grow older would be unknown, or at best very minor inconveniences, if we simply left the scene when nature intended.

Another dilemma for the immune system lies in the way it was designed. In applying its force, the immune system is essentially blind. With a few useful exceptions, it has no way of knowing whether a microbe that has invaded the body, and possibly taken up residence inside a cell, is potentially pathogenic or completely harmless. It simply knows the microbe does not belong there, and will relentlessly, blindly pursue it until either it is cleared from the system or until, in extreme cases, the immune response finally destroys the host.

What about autoimmunity? Where does *that* come from? It could be viewed as just another way nature has of being sure we don't hang around too long, using up valuable resources. But in fact, with a few exceptions most autoimmune diseases are not all that life-threatening. They make life miserable, but they don't usually kill us. So how do they fit into the grand scheme of things? Why does the immune system turn against itself?

Although many autoimmune diseases seem almost certainly to

represent an unprovoked attack of the immune system on self, the possibility that at least some such diseases are due to cryptic microorganisms continues to intrigue many immunologists. If even tiny traces of invading microbes remain lodged in human tissues after an infection, they argue, immune-based disease could ensue. Although the microorganisms would be present in amounts too low to be detected by even the most sensitive clinical tests, they would still be detected by the immune system. In such small amounts, even the most virulent microbes would themselves be unlikely to cause disease, but the attempts of the immune system to ferret them out and destroy them could cause extensive damage to apparently normal human tissues. The problem with such hypotheses, of course, is that they are virtually impossible either to prove or to disprove, since they posit things that cannot be measured.

An interesting variant of this hypothesis is something called *antigenic mimickry*. What if an invading bacterium or virus contained a protein, a very small region of which was identical to some human protein? The odds against this are by no means astronomical. In the process of responding immunologically to that particular stretch of the foreign bacterial or viral protein, might we produce antibodies or activated T cells capable of attacking the corresponding human protein? There is intriguing evidence that this might actually happen, and in a few cases the evidence is quite strong that it does. For example, rheumatic carditis is an autoimmune condition in which we produce antibodies against our own heart proteins. This disease almost always follows on the heels of a previous infection with streptococcal bacteria. Although the antibodies causing the damage are clearly directed against human heart muscle proteins, it has been suspected for years that the antigen triggering the antibodies was actually streptococcal in origin. Scientists have now isolated a small segment of one of the surface proteins of streptococcal bacteria that induces the antibodies that cross-react with human heart muscle.

So quite likely some diseases that we think of as autoimmune may be various forms of spillovers from normal immune attacks against foreign invaders. But equally likely we may just have to come to grips with the possibility that the immune system does, on occasion, decide to attack self, unprovoked by outside agents. Is this simply one more cross we must bear, one more price we must pay for an immune system that does a pretty good job most of the time? Or could it be that autoimmunity is a normal part of human biology, playing a more profound role than malicious aggravation?

A close pursuit of this very question has led to some intriguing insights into how the immune system is put together. For example, it has been observed that the immune system, both in terms of T cells and of B cells, seems to be directed, right around the time of birth, largely against self. If we examine the antibodies in the blood of human infants just after birth, we find that a rather high percentage of them are directed at self antigens. This condition disappears a short time after birth, but it is as if, just prior to that instant when the immune system was taking that last look around to define "self" at birth, it was actually using self antigens to prime itself, to get itself up and going. This phenomenon is thus probably connected to the issues of tolerance and fetal development discussed earlier in this chapter; the immune system is busy investigating what is and is not self. As far as we can tell, this self-reactivity causes no harm, either in the fetus or in the newborn. But beyond being simply a neutral phenomenon, this observation has prodded scientists to wonder whether in fact this mild form of self-reactivity by the immune system may actually be a necessary and beneficial step in the development of the fetus. Thus, both at the very beginning and the very end of life, we see significant levels of self-reactivity by the immune system. Right now, no one knows what that means, but you can be sure it is a question that will continue to be pursued.

SIX

When the Wall Comes Tumbling Down: AIDS

Primary immune deficiencies—like Bruton's XLA and the SCID disorder that afflicted David the Bubble Boy—occur exclusively in children. They arise from one source only: the inheritance of defective genes controlling some critical aspect of immune responsiveness. These diseases provide a dramatic form of natural evidence that the immune system, despite its faults and problems, is absolutely essential to human survival. And if primary immune deficiencies do not make that point clearly enough, we have further evidence in the form of secondary, or acquired, immune deficiencies.

Secondary immune deficiency diseases, which are much more common clinically, are mostly an adult problem, although they can occur in children. They arise from two sources, neither of which is inherited; one is manmade and the other is natural. Manmade causes have at least in the past been the most common, and they include such things as some of the treatments used for cancer and for organ transplantation. In both cases, the treatments used may result in *immunosuppression*. In cancer, the object of both radiation therapy and chemotherapy is to destroy rapidly dividing cancer cells, to keep them from spreading. Unfortunately, cells of the immune system, and in particular the stem cells residing in the bone marrow, are exquisitely sensitive to both these treatments. It would be much easier to rid the body of cancer cells if higher doses of cancer-fighting radiation or drugs could be used. But well short of such doses, the body's immune system begins to collapse from the cancer treatments, and symptoms of acquired immune deficiency begin to appear.

A similar problem arises in organ transplantation. Here, immunosuppression is not a side effect of treatment—it is the object, the purpose of treatment. Physicians specifically set out to suppress the immune system so that the incoming heart or kidney won't be rejected. But this is a delicate and hazardous balancing act. The immunosuppression needed to allow the new organ to survive also leaves the body open to a wide range of infections; the treatment required to prevent rejection induces a very real *acquired immune deficiency*. In the early days of organ transplantation, the resulting infections, rather than failure of the transplanted organ, were a leading cause of death.

The most prominent example of a secondary or acquired immune deficiency stemming from natural causes is of course AIDS—the acquired immune deficiency syndrome. As is now well known, AIDS arises when a virus attacks and destroys key elements in the immune system, leaving the victim every bit as vulnerable to a wide range of pathogens as a child with SCID. Viral infections are perfectly natural processes, but when the virus attacks the very system meant to protect the body against, among other things, viruses—the results can be particularly devastating.

AIDS was first reported in June 1981 by Michael Gottlieb, a young physician at the University of California, Los Angeles, who noted five cases of *Pneumocystis carinii* pneumonia in five previously healthy men who were homosexual. *Pneumocystis carinii* is what is known as an *opportunistic pathogen*—a disease-causing microbe that can be found in small numbers in many normal individuals, but which grows out to dangerous proportions only in those whose immune systems have somehow been compromised—like transplant or cancer patients. Gottlieb was familiar with *P. carinii* in these settings but was highly surprised to find this form of pneumonia in a cluster of five men who had no apparent reason to be immunosuppressed. In fact, he found this situation so striking that he reported it to the Centers for Disease Control (CDC) in Atlanta. The CDC is a federal agency that tracks diseases as they come and go throughout the United States. Barely one month later it was reported that twenty-six homosexual patients seen during a short time period in another state had *P. carinii* pneumonia and/or Kaposi's sarcoma, a rare form of skin cancer. In less than a year, new cases obviously belonging to this syndrome numbered in the hundreds. Thus began a journey into the unknown, a journey whose end is still beyond our vision. Unlike primary immune diseases, AIDS would not teach us anything we didn't already know about the immune system. It *would* highlight with blinding clarity the central role of the CD4 helper T cell in the immune system, but we already knew that. We did not need to lose tens of thousands of human lives to get the picture.

AIDS is a disease caused by a *retrovirus* called the *human immunodeficiency virus*, or *HIV*. Where did HIV come from? Where has it been throughout human history? Why have we never seen it before? Why is it only now, at the end of one of the most productive centuries ever in human medicine, wreaking havoc among human populations? These are questions asked by the lay public and the health establishment alike. We don't know all the answers to these questions, but a reasonably good picture is beginning to emerge from the void.

It now seems likely that HIV has been around for a very long time, but in a slightly different form, called SIV—*simian immunodeficiency virus*. SIV appears to be a relatively benign virus of monkeys and other primates in Africa. It is benign because over the ages it has learned to live with and profit from its natural host without killing it. The ideal parasite–host relationship is one in which the parasite takes as much as it can from the host, but always allowing the host to live so it can continue to support the parasite. Sometime in the relatively recent past—probably in this century—continuing human encroachment on the African ecosphere resulted in the jump of this virus from monkeys and apes to humans. Passage from lower primates to humans probably occurred through a bite; saliva is known to contain this virus in both humans and lower primates. The virus appears to have mutated wildly in humans, maintaining its basic character as a retrovirus, but differing now markedly in its HIV form from the original SIV form as it tried to adjust to its new environment. Unfortunately, HIV has not yet learned to live in harmony with humans as a benign parasite; it has caused one of the most deadly diseases seen on this earth since smallpox reigned supreme in the Middle Ages.

When HIV was identified in 1984 as the pathogen responsible for AIDS, everyone breathed a deep sign of relief. It was a virus of a generally recognized type; virologists already knew a fair amount about retroviruses. Surely a means for stopping or at least controlling this disease must be just around the corner. But only months later came a heart-stopping announcement: Eighteen of the first nineteen HIV isolates taken from AIDS patients were antigenically different. What that meant to immunologists was that the coat porteins in which HIV wraps itself, the parts of the virus that are detected by the immune system, must be mutating at an extraordinarily high rate. Was this a reflection of the virus's attempt to adapt to its new host? Possibly, but no one knew for sure. In practical terms it meant that it would be extremely difficult to prepare a vaccine against HIV. Any form of the virus killed and used as a vaccine one month would probably induce a perfectly good immune defense against that strain of the virus, but the resulting defense would be perfectly useless against forms of the virus floating around the next month.

Unfortunately, this has proved all too true. The same is true of colds caused by the flu virus. Influenza virus also changes its coat proteins at a fast rate. That is why there is no vaccine that can protect us against all forms of the flu, and why we never build up immunological memory from one cold that can protect us against the next wave of flu virus. No one has ever tried very hard to solve this problem, because usually our own immune systems manage to get on top of each influenza infection and rid our bodies of it. But this doesn't happen in AIDS, for reasons we shall explore in this chapter.

AIDS as a Clinical Problem

How big a medical problem is AIDS? The World Health Organization now estimates that twenty million people worldwide have been infected by the AIDS virus. Half of these are in sub-Saharan Africa, but the rate of increase of AIDS in Asian countries may soon put that region on an equal footing. In the United States some three hundred thousand cases have been diagnosed to date; two-thirds of these have already died, and an estimated two million more people may be infected with the AIDS virus. In more than sixty U.S. cities, AIDS is now the leading cause of death among males in the twenty- to fifty-year-old age bracket. In California, it accounts for 24 percent of all deaths among males between twenty-five and forty-four years of age. Males in this bracket account for 54 percent of the work force nationwide.

How do we make sense of figures like these? What sort of perspective can we put them into? On the one hand, AIDS still does not really come close to something like smallpox or hepatitis B as a killer of human beings in a historical sense. Nor, for that matter, can it yet compare with the influenza epidemic that swept thirty million human beings from the face of this earth just after World War I. What is so frightening about AIDS is the speed with which it is spreading, the incredible rate of increase in the number of cases diagnosed each year, with absolutely no cure in sight. One can forgive medicine for not dealing with plagues and epidemics in the past, but this is the age of computers and fiber optics and humans long ago on the moon. Why can't we get on top of this thing? Where will it end?

We don't know yet where it will end, but we should probably dig in for the long haul. This is a problem that is going to occupy a major proportion of the world's scientists and health care professionals for the foreseeable future, and it is going to require that we all develop a reasonable understanding of just what it is we are facing. Thus, what follows here is an introduction to AIDS, a primer if you will. We will probably all have read several books about this malady before it is over.

First, AIDS is not a disease in the sense we normally use that word. It is, as its name states, a *syndrome*—a collection of individual disease symptoms seen together frequently enough that they are assumed to have a common underlying cause. Almost everyone now agrees that the causative agent in AIDS is the virus HIV. But HIV as a pathogen does not itself directly cause most of the disease symptoms associated with AIDS. Rather, it cripples the immune system, by means we will look at in a moment. The opportunistic pathogens that are able to thrive in the absence of an effective immune response are what actually cause the diseases seen in the clinic. Different AIDS patients may display disease symptoms triggered by one or another of these pathogens, or several simultaneously, or they may suffer from one of the cancers induced by HIV, alone or in combination with one or more of the opportunistic diseases.

There are several stages in the progress of an HIV infection recognizable by medical workers at most AIDS centers. These stages have been organized and defined by the CDC in Atlanta, and they are shown in Table 6.1.

The acute illness stage refers to the first set of symptoms indicating that the body has been invaded by the AIDS virus. These symptoms may appear anywhere from several days to several months after infection by HIV. The symptoms are not unlike other viral infections; the body's initial response to HIV is much the same as it is to any other virus. There may be fever, achiness, sore throat, and other flu-like symptoms. There are in fact few outward symptoms at this stage that would specifically point to infection with HIV. Only if the patient belonged to an identifiable high-risk group for AIDS might a physician carry out the necessary follow-up procedures to look for evidence of HIV. And there would be no problem finding it; the blood of patients at this stage is loaded with mature HIV particles, which rapidly spread throughout the body.

Within a month or two after disappearance of the symptoms just described, an event called *seroconversion* takes place that further signals the presence of HIV in the body and defines entry

Table 6.1. The Stages of HIV Infection Leading to AIDS

Stages of HIV infection as defined by the Centers for Disease Control	Number of CD4 T cells/mm³ of blood	Opportunistic infections seen
Preinfection	1,000	None
I. Acute illness; seroconversion	1,000	None
II. Chronic asymptomatic infection	800–500	Tuberculosis
III. Lymphadenopathy	500–200	Herpes I; candida
IV. ARC and AIDS Pneumocystis, opportunistic infections, AIDS-related tumors, neurological disorders	200 and below	Histoplasma; *P. carinii*; Herpes II; cytomegalovirus (CMV); Kaposi's sarcoma

into Stage II. Seroconversion means the appearance in the blood of antibodies produced by the immune system's B cells that recognize HIV. Current evidence suggests that seroconversion, if confirmed by other tests for HIV, marks the unequivocal onset of a process that, in virtually all infected individuals, will end in the disease AIDS.

But this process usually takes a long time, presently about ten years for most patients. During much of this time, the infected individual remains in Stage II—the asymptomatic stage—showing few if any outward signs of infection, although almost certainly some degeneration of the immune system is occurring. Antibodies to HIV remain present throughout Stage II, although only a few HIV particles can be found in the blood, or in cells that circulate in the blood.

Stage III signals the approach of more serious disease. *Lymphadenopathy*—literally lymph gland abnormality—is characterized by swollen and tender lymph nodes in the neck, underarm area, groin, and other locations throughout the body. This stage can last for several months, and some of the symptoms from the earlier acute stage may return: fever and chills, tenderness in many areas of the body, chronic fatigue. Stage III merges gradually into Stage IV, a complex of conditions moving through what was formerly referred to as ARC (*AIDS-related complex*), and on into full-blown AIDS. Prior to Stage IV, patients are referred to simply as HIV-infected, or HIV-positive. When they begin to express the specific set of disease symptoms defining Stage IV, they are formally described as having AIDS. The average time to death once a patient enters Stage IV is one to two years.

As we will see in this chapter, HIV causes a gradual loss of CD4 helper T cells from the blood circulation of infected individuals. When the level of circulating CD4 T cells falls below roughly half the normal value, HIV-infected individuals begin experiencing infection both by external pathogens and by opportunistic pathogens already present in the body but controlled directly or indirectly by the CD4 T cells. The correlation of CD4 T-cell levels and the appearance of these opportunistic pathogens with the CDC staging scheme is also shown in Table 6.1.

Although the disorders emerging in Stage IV of an HIV infection are incredibly complex, and not the same for all AIDS patients, they fall into three major categories: opportunistic infections, AIDS-associated cancers, and neurological disorders.

Opportunistic Infections. Opportunistic infections arise from potential pathogens that are already within us. As a result of some previous exposure to the pathogen, we probaby experienced some mild symptoms of disease, after which our immune systems got on top of the infection but did not completely clear it from the body. These potential disease-causing organisms usually hide in some particular tissue or cell type where they are relatively safe from the immune system, but whenever they try to come out, they are picked off in a hurry. It is only when the immune system is compromised by some other problem that these pathogens can roam free in the body and wreak havoc unopposed.

The opportunistic infections seen in AIDS are similar to those seen in other secondary immune deficiencies, such as those caused by treatments for cancer, or those seen in immunosuppressed organ transplant patients. And they are every bit as deadly: The majority of deaths in AIDS patients are still due to opportunistic infections. Funguses are a particular problem. *Histoplasmosis* is a set of disease symptoms caused by the fungus *Histoplasma capsulatum*. *Histoplasma* is quite common in portions of the Midwest, but rare on the East and West coasts. Because most of the early AIDS cases showed up in large coastal metropolitan areas, it took some time to realize that histoplasmosis was surfacing as an AIDS-associated opportunistic infection. The initial symptoms are rather nonspecific—fever, respiratory distress, weight loss—and are similar to those caused by many other opportunistic and external pathogens. As the infection progresses, internal organs such as the spleen, liver, intestines, and lymph nodes become involved, and if the infection is not checked it can be fatal. Fortunately, *Histoplasma* is very sensitive to the drug amphotericin B, so if it is diagnosed early and correctly, the infection can usually be brought under control.

Another fungus commonly seen in AIDS patients is *Candida albicans*, which is responsible for a condition called *thrush*, in which the fungus begins to coat the tongue and membranes of the mouth. *Candida albicans* is also commonly present in healthy people. In humans, the *Candida* fungi live in the gut and the genital area but are normally kept in check by bacteria that co-inhabit the same locations. A major cause of candida outbreaks in AIDS patients is the administration of potent antibiotics needed to fight off infections that occur in the absence of a functioning immune system. These antibiotics cripple the bacteria that keep *Candida* under control, leading to outbreaks of *Candida* at many places in the body.

By far the deadliest of the opportunistic fungi is *Pneumocystis carinii*, which is not well understood as a biological organism. Until a few years ago, it was thought to be a parasite rather than a fungus. Like *C. albicans*, *P. carinii* can be found in most healthy people, largely in the lungs; like *Histoplasma*, it is kept under control by the body's immune system. In AIDS patients, as in transplant or cancer patients who are immunosuppressed by drugs, *P. carinii* causes a particularly deadly form of pneumonia that is difficult to treat. This pneumonia develops in about 80 percent of AIDS patients. Initially it was a major cause of death, and it still remains a serious problem, but it has been brought under at least some degree of control by drugs like pentamidine, which can be sprayed directly into the lungs. In fact, improved management of fungal infections in general has led to an interesting shift in the infectious-disease spectrum seen in AIDS patients. The majority of infectious-disease problems are now from standard pathogens normally living outside the body, rather than from opportunistic infections by endogenous pathogens. Nevertheless, in terms of mortality, opportunistic fungi remain a serious problem.

In addition, AIDS patients suffer from a number of opportunistic viruses that are relatively harmless to the population as a whole, such as *Herpes simplex* (both Types I and II), and *cytomegalovirus*. Type I herpes is the virus that causes common cold sores. The virus lives in nerve cells that make up the nerve fibers leading from the brain to the face. From time to time the virus makes its way down these nerve fibers and attacks the mucous linings around the nose and mouth. It can occasionally affect the eyes as well. In persons with a healthy immune system, T cells keep these herpes outbreaks limited to usually no more than a small sore at one of these sites that resolves in a few days. In AIDS patients, when the number of CD4 T cells in the blood drops to a low level, herpes Type I lesions may become massive, covering much of the head and neck region. Not only is this extremely painful but it may lead to a type of pneumonia, and the open sores may become infected with a wide range of environmental pathogens. If the lesions involve the eye, blindness can result.

Type II herpes is a sexually transmitted disease that in most people results in occasional small sores on the genitalia. Again, this is a relatively minor problem in someone with an intact immune system, but one that can blossom into major problems in an AIDS patient. *Cytomegalovirus* (CMV) is related to the herpes family of virus. Approximately half of the U.S. population is infected with CMV, but people with healthy immune systems are rarely symptomatic. In AIDS patients, as with immunosuppressed transplant or cancer patients, CMV causes a wide range of serious health problems, including a type of pneumonia as well as ulcers, brain and liver damage, and blindness.

Finally, we should probably add tuberculosis to the list of AIDS-associated opportunistic infections. Tuberculosis (TB) seemed on the verge of disappearing completely until the early 1980s, when the number of cases suddenly began to rise—almost entirely in AIDS patients. Many healthy people harbor latent TB bacilli, but as with other opportunistic pathogens these are kept under control by the immune system. Yet TB is not usually seen in acquired immune deficiencies arising in, for example, transplant or cancer patients who are immunosuppressed. For some as-yet-unknown reason, TB is becoming a serious problem in AIDS patients, and it is clear that in a great many of these cases we are seeing the reemergence of latent disease. Moreover, the strain of TB that is emerging is very aggressive and resistant to many of the drugs normally used to treat this disease. There is great concern that this virulent form of TB could spread into the general population, setting us back fifty years in the fight to eradicate this crippling malady.

Thus we see that AIDS is not a single disease; it is many diseases, caused by many different pathogens, some of which come from the outside, and some that live within us. Each of these diseases underlies a different dysfunction; each requires a unique treatment, and very often AIDS patients may have more than one disease ravaging their body at the same time. As we will see, this is only one of many factors complicating the treatment of AIDS.

AIDS-Associated Cancers. To the extent that the immune system is involved in routine surveillance and suppression or elimination of cancerous cells, it might be expected that destruction of the immune system would be accompanied by the development of tumors. To some extent this is indeed seen in AIDS, but at least in the past most AIDS patients have died of other causes before many forms of cancer would have time to become a clinical problem. However, as treatments for some of the opportunistic infections associated with AIDS have gradually improved, the life span of AIDS patients from the time of diagnosis has increased slightly, and will no doubt be extended even further in the coming years. Thus, paradoxically, we may anticipate that cancer will become a more serious complication of AIDS as death rates from other causes drop.

The most common cancer seen in AIDS patients is Kaposi's sarcoma. It was one of the first clinical syndromes used to define AIDS, and in the early years of the AIDS epidemic it was a major disorder seen in AIDS patients. Kaposi's sarcoma is an unusual cancer in many ways, not the least of which is that some scientists are not entirely sure it is a cancer. Many think it may be caused by an opportunistic pathogen. But for the time being at least it is still listed as a cancer, and we will refer to it as such here.

Kaposi's sarcoma (KS) was known long before the advent of AIDS—it was first described in 1877—but this "classic form" of KS is somewhat different from the form showing up in AIDS patients. Classic KS was seen mostly in elderly males from Mediterranean cultures, and to some extent in Africans. It is a very slow-growing cancer, mostly confined to the skin. It can be fatal, but in fact is rarely listed as a cause of death.

However, AIDS patients have a much more aggressive form of KS. It is normally detected first as purplish-brown skin lesions on the surface of the body or around and in the mouth, but it spreads rapidly throughout internal organs as well. Kaposi's sarcoma was apparent from the earliest days of AIDS, when perhaps 80 percent of AIDS patients were afflicted with it. For reasons not presently understood, that figure has fallen to about 20 percent. It is possible that some of the treatments targeted to opportunistic infections, or against the virus itself, are having an unexpected effect on the development of KS. Although many AIDS patients die with active KS, it is not obvious that KS is a cause of death. Nevertheless, the development of KS in an HIV-positive individual is a clear sign that Stage IV has begun, and that the prognosis is not good.

In addition, AIDS patients experience a higher than normal risk for lymphomas (tumors of the immune system), especially B-cell lymphomas. In a B-cell lymphoma, a particular clone of B cells grows extremely rapidly until it fills virtually all the B-cell compartments in the body, crowding out other B cells that might be needed to make various different antibodies. This is certainly the last thing someone who is already immunocompromised needs! The Epstein-Barr virus (EBV), which is present in many healthy individuals but rarely causes anything more serious than mononucleosis, is a common cause of lymphoma in AIDS patients. You may remember that it was EBV lymphoma that complicated the bone marrow transplant in David, the "Bubble Boy." B-cell lymphomas associated with AIDS are very aggressive and hard to control. They are aided in their development by the gradual loss of T-cell function in AIDS patients. The body's T cells normally detect and eliminate these types of tumors.

At present, about 3 percent of AIDS patients need active treatment for some form of cancer other than KS. However, based on the small but significant increase in life expectancy for AIDS patients after diagnosis, epidemiologists anticipate a similar gradual rise in patients with non-KS cancer. It is expected that over the next several years, there will be at least five thousand new cases of non-KS cancer annually in AIDS patients. These cancers will probably continue to be very difficult to treat. Strategies that doctors might use to treat a particular type of cancer in most patients do not always apply to AIDS patients, who are already very ill and cannot tolerate many standard cancer treatments. Moreover, AIDS patients may already be taking drugs for opportunistic infections that are incompatible with chemotherapeutic drugs. As we will see shortly, this often leads to agonizing choices for AIDS patients and their physicians.

AIDS-Associated Neurological Disorders. By no means does AIDS lack in tragic dimensions, but surely one of its sadder aspects is the prevalence of disorders of the nervous system, particularly dementia, seen in a significant proportion of the AIDS patient population. Eighty percent or more of AIDS patients show evidence of damage to the brain or nervous system at autopsy. About one-third of AIDS patients display overt neurological abnormalities while alive, and in one-tenth the symptoms are serious enough to be disabling. It was several years after AIDS was first described before these HIV-induced disorders could be sorted out from the depression that might be expected to accompany *any* catastrophic illness, or damage to the nervous system caused by opportunistic or other infections.

The most serious neurological and psychiatric disturbances are grouped together in something called *AIDS dementia complex*, or *ADC*, which includes elements of dementia, impaired motor functions, and behavioral (personality) changes. Early symptoms involve loss of memory, inability to concentrate on simple thoughts and tasks, and a general "slowness" in thinking. This is often accompanied by difficulty in coordinating hand–eye functions, which may progress to balance–coordination problems, and increasing loss of the ability to move about. In advanced cases, the patient may enter a near-vegetative state, with minimal intellectual or social comprehension, and loss of the most basic body functions.

It is now clear that HIV directly infects certain macrophage-like cells in the brain called *microglial cells*. It is thought that HIV may be brought into the brain by roving macrophages, which then somehow transmit the virus to the microglial cells. The HIV does not kill these cells; in fact, the virus doesn't even replicate terribly well in them. Interestingly, HIV does not appear to infect the working cells of the brain, the *neurons*, which carry nerve impulses between different parts of the brain, and between the brain and the rest of the body. Yet at autopsy, structures of the brain composed of neurons show severe damage and disarrangement. How might this happen?

Studies suggest that one possibility is that viral proteins released by infected cells may be directly toxic for nerve cells. This can be seen in cultures of neurons exposed to HIV and its component proteins. But it is also highly likely that nonviral proteins released by both T cells and macrophages may contribute to nervous system damage. Whenever there is an infection of any sort, both T cells and macrophages produce proteins that make their way to the brain to let it know there is a problem. These cytokines can definitely alter brain function, usually in a way that helps the body fight infection. For example, a protein produced by macrophages involved in fighting an infection, called interleukin-3 (IL-3), tells the brain to turn up body temperature—to produce fever. These cytokines are normally produced in response to a specific problem, and the signal to the brain disappears as the problem is brought under control and cytokine production ceases.

But in a chronic infection such as that produced by HIV, where T cells and macrophages are themselves infected, this chemical communication may go terribly awry. All sorts of immune-system cytokines, perhaps carrying conflicting chemical messages, are produced in enormous quantities and released into the bloodstream from sites throughout the body. This chemical barrage may produce a type of information overload, stressing the brain beyond endurance as it tries in vain to sort out and respond to the various messages. Some of these messages almost certainly come from the brain-associated cells that carry HIV, but it is equally likely that CD4 T cells and macrophages throughout the body contribute to this chemical cacophony, literally leading to a kind of brain "meltdown."

AIDS as a Problem in Virology

The human immunodeficiency virus may now be the most intensely studied virus on the face of the earth. We need to know every single aspect of how this virus infects cells, reproduces, and leaves dying cells to start new infections if we are ever to find its Achilles' heel and disable it for good. We need to know every single gene, every protein in its structure. We need to know what parts of the infected cell it uses to help it make more copies of itself and to devise ways to deprive it of this help. Knowledge of this type does not come cheaply, it takes time, it takes money, and it takes commitment. It takes being in it for the long haul.

Interestingly, HIV—this most deadly of human viruses—is rather fragile as viruses go. When my own lab first joined AIDS researchers at UCLA to test a possible antiviral agent on HIV, we were surprised at how difficult it was just to keep HIV propagating in human cell cultures. Great care had to be taken not to damage the virus during handling, or else it would lose its ability to infect and replicate in human cells.

It is known that HIV is an RNA retrovirus, which means that its genetic blueprint is written in the RNA code, rather than in the DNA code used by all animal (including human) cells. The entire virus consists simply of this piece of RNA wrapped in a small number of *coat proteins*. Like all viruses, HIV does not have the machinery necessary for reproducing itself; to do so it must infect a living cell and exploit that cell's materials and energy to make more virus. The first step in the infectious process is thus binding to a living cell. One of the prominent proteins making up the coat of HIV is a *glycoprotein* (a protein that contains sugar molecules in its structure) called *gp120* (*gp* is an abbreviation for glycoprotein; 120 refers to its size in atomic units). HIV uses gp120 to bind to the cell it is going to infect. The gp120 protein specifically recognizes and binds to the CD4 molecule found mostly on CD4 T cells, but to a lesser extent on macrophages and possibly certain brain cells. As we will see, it is this predilection of HIV to bind CD4 molecules that ultimately makes this virus so deadly.

Once bound to the outside of a cell, the HIV particle crosses the cell membrane and enters the cytoplasm, shedding its entire protein coat in the process. HIV brings along with it information for producing an enzyme called *reverse transcriptase*, which allows it to transcribe its RNA into DNA, the genetic language of the host cell. This step is partly responsible for the high mutation rate seen in HIV. Whether by chance or design, large numbers of mistakes are made during reverse transcription, leading to a high rate of mutation of the resulting DNA. Most of these mutations are likely to be deleterious to the virus, but that doesn't matter. The virus reproduces so rapidly inside a living cell that errors are affordable, as long as a few functional viruses are made in the process. The tremendous advantage to the virus is that on rare occasions these mutations will produce a new strain of virus that is even more effective than the virus that originally infected the cell. Mutations in the coat proteins of the virus may be particularly important in helping the virus escape destruction by the immune system.

The HIV DNA copied from the infecting virus RNA makes its way into the nucleus of its new host, where it inserts into one of the host chromosomes. In this form, the HIV DNA is referred to as a *provirus*. This is a particularly insidious event, because from that point on the host cell regards the HIV proviral DNA as part of its own DNA, and will follow whatever instructions are encoded therein. But first there is a period of quiescence, or *dormancy*, in which the proviral DNA just sits in the host DNA, biding its time, sending out no instructions. This period ranges from days to months; it is the time between the moment of infection with the virus and the "acute illness" stage (Stage I) shown in Table 6.1. During this time the unsuspecting individual harboring HIV is entirely asymptomatic. There are no mature HIV particles or gp120 proteins floating around in the bloodstream at this stage, nor yet any host antibodies directed to any part of HIV; the infected individual is seronegative.

No one really knows what determines whether or how long the provirus will remain dormant, or what factors cause it to suddenly become active. Activation of the HIV proviral DNA means that it starts sending messages (in the form of messenger RNA) out of the host cell nucleus and into the cell's cytoplasm, where proteins are made. These messages contain instructions for building new HIV components that will assemble into thousands of particles of the same deadly virus. Once this process is up to speed, nearly half of all the molecules made by the infected cell are directed by viral DNA and used to make new virus. Tens of thousands of new copies of the virus may be made and released by the cell. The end result of this process is that the infected cell dies. Once HIV particles begin to be released into the bloodstream, an initial wave of HIV antibodies is produced by the host, and these can now be detected in the blood (seroconversion).

There is one very curious observation related to CD4 cell death that has been confirmed by virtually every lab that has studied this problem. Even at the very peak of active AIDS, HIV genes are actually expressed in less than one in a thousand circulating CD4 T cells! This led Peter Duesberg, an eminent virologist at the University of California at Berkeley, to suggest in the late 1980s that HIV might not be the causative agent in AIDS. He correctly asserted that this failure violated one of the basic tenets of microbiology—one of the rules established almost a century ago by Robert Koch—that in order to implicate definitively a pathogen in a specific disease, it is necessary to show that the suspected pathogen can in fact be isolated from an animal or person with that disease. Duesberg proposed instead that the collection of problems identified as AIDS is due to lifestyle-associated behavior such as drugs, alcohol, malnutrition, and other forms of self-abuse that leave the body vulnerable to opportunistic and environmental pathogens—one of which is HIV.

In the face of the tremendous amount of evidence that HIV *is* the causative agent in AIDS, Duesberg's proposal caused considerable consternation, but given his outstanding scientific reputation it could not be lightly dismissed. However, in the last several years we have learned more about the details of the life history of HIV, and at least one part of the mystery has finally been resolved. The problem was that almost all of the previous attempts to look for traces of HIV in CD4 T cells were carried out using T cells from blood, which are easy to obtain. When the CD4 T cells living in lymph nodes and intestines of AIDS patients were finally examined, they were found to contain very large amounts of HIV DNA and even RNA. Moreover, one could see abundant mature virus particles within the lymph nodes, sticking to dendritic cells and macrophages. So now we can safely say that the infectious agent *can* be found in AIDS patients, and that that agent is in fact HIV. It is not seen in the circulation because the lymph nodes are doing exactly what they were designed to do—filtering out infectious agents. Unfortunately, this also has the effect of exposing CD4 T cells in the lymph node (where 98 percent of them may be found at any given time) to HIV. As many as 25 percent of CD4 T cells in the lymph nodes or gut tissues of infected patients may contain HIV DNA. This clearly is a major site for interaction of HIV with the immune system. Thus, HIV is able to subvert the normal filtering function of lymph nodes and redirect this function to actually enhance its own chances of finding a CD4 T cell to infect. It seems rather amazing in retrospect that this fact went unrecognized for over ten years.

This recent finding of HIV expression in lymph node T cells may be related to another perplexing and unresolved question about AIDS and HIV infection: the prolonged course of this disease. Stage II of HIV infection is the so-called chronic asymptomatic phase. This period can last anywhere from three to eight years, during which the patient is for all practical purposes completely normal—truly asymptomatic. This is very unusual for human diseases caused by infectious agents. For several years after discovery of the AIDS virus, it was thought that during Stage II of infection the virus was in a dormant state—still there, integrated into the host DNA, but not doing anything active, like replicating. Now we know that this is not true. All available evidence suggests that during this period the virus is doing everything it can to break loose from the lymph node environment where it is trapped and to destroy the host, but it is kept in check by the immune system. During Stage II, although it is hard to find HIV DNA present in circulating CD4 T cells, HIV particles *can* be isolated from the serum of infected individuals. Moreover, it is during this period that we see the appearance of new genetic variants of HIV—the virus is constantly changing its coat, further confounding the immune system that is trying to stop it.

All in all, the evidence that HIV causes AIDS must now be considered overwhelming, if largely circumstantial. It will have to remain circumstantial; no one is going to propose testing Koch's postulates directly by deliberately exposing human beings to HIV under controlled laboratory conditions. Yet several tragic experiments have occurred in unintended and utterly uncontrolled ways. Two of these provide the strongest possible direct evidence that HIV causes AIDS. In the first instance, hemophiliacs who were inadvertently exposed to HIV through administration of contaminated blood and blood products, prior to the establishment of rigorous screening procedures, show an extremely strong correlation of the disease AIDS with exposure to HIV. In a second demonstration, three laboratory workers accidentally exposed to a highly purified strain of HIV have all come down with symptoms of AIDS. All three have now developed very low CD4 T cell levels, One has developed *P. carinii* pneumonia and has been formally diagnosed with AIDS. In the face of these two accidental "experiments," few would continue to argue that HIV does not cause AIDS. To suspend efforts to educate the public to the fact that it does would be unconscionable.

The Immunology of AIDS

So how does all of this lead to AIDS? Why does infection with HIV, alone among all viral infections, result in an acquired immune deficiency? In terms of the human immune system, the single most important fact about HIV is its target in the human host. We have seen that the gp120 coat protein binds selectively to the CD4 molecule, the surface protein that distinguishes the CD4 helper T cell. As a result, HIV selectively infects and ultimately destroys human T helper cells. From a clinical point of view, it has been clear from the start that the most reliable predictor for the progression of AIDS as a disease is the level of viable CD4 T cells remaining in the blood. It is almost possible to correlate the stages of HIV infection presented in Table 6.1 with the level of CD4 T cells remaining in the blood. Most HIV-infected individuals with CD4 counts above 500 (500 cells/mm^3 of blood), although seropositive, are still asymptomatic. From the time of HIV seroconversion to a CD4 count of 500 takes about four years. Between a count of 500 and 250, oral candidiasis (a fungal infection of the mouth) and tuberculosis are the most common problems; at 200 to 150 (about eight to ten years after seroconversion) it is Kaposi's sarcoma and lymphoma that are seen most frequently; below 150, deadly opportunistic pathogens such as *P. carinii* and cytomegalovirus make their appearance.

The consequences of CD4 T-cell depletion are incredibly complex, for CD4 T cells affect virtually every phase of our immune responsiveness. And not just antigen-specific components such as other T-cell subsets and B cells. The CD4 T cells, through the cytokines they produce, affect macrophages, dendritic cells, granulocytes; even bone marrow and thymus, the "master organs" of the immune system. It is believed that CD4 T-cell products are involved in communication between the immune system and the brain. In one's wildest imagination, one could not possibly pick a worse cell to serve as the target for an infectious virus. Either the virus itself will kill or disable the CD4 T cell, or the immune system will sense the presence of the virus within the CD4 T cell, and destroy it. Either way, we lose. AIDS is in that sense the adult equivalent of SCID; both the T- and B-cell arms of the immune system are wiped out. But there is one respect in which AIDS is even worse. In SCID, there is always the possibility of a bone marrow transplant. The success rate is not the greatest, but with a good tissue match there is a reasonable fighting chance. In AIDS, however, a bone marrow transplant would be of no value whatsoever. Because HIV also infects but does not kill cells such as macrophages and dendritic cells, these cells serve as reservoirs of infectious virus in affected individuals. They will continue producing HIV for the life of the patient, regardless of what happens to the T cells. New CD4 T cells growing out from a bone marrow transplant would become infected within minutes of their emergence.

The ability of HIV to infect macrophages, by the way, should not be passed over lightly. In addition to acting as a reservoir for HIV, macrophages may also transport HIV into parts of the body the virus might not otherwise reach, such as the nervous system. Macrophages have been considered by some in this respect to be the cellular equivalent of a Trojan horse. Moreover, when macrophages are infected with HIV, as with any other pathogen, they produce chemicals that can lead to *cachexia*, a state characterized by excessive weight loss. In fact, macrophage infection with HIV, rather than CD4 T-cell infection, is generally considered to be the major factor in the wasting syndrome accompanying AIDS.

Strangely enough, to this day no one knows exactly how HIV kills CD4 T cells. We do know that most strains of HIV kill CD4 T cells directly; that is, they are *cytotoxic*. CD4 T cells incubated in a test tube with most strains of HIV will die in the absence of any other agent. The simplest possibility would be that the process of newly made viruses bursting out of the cell is lethal for the cell—tearing up its membrane, creating irreversible damage to the cell's innards. This happens in many viral infections but is not true for HIV. Only cells with high concentrations of CD4 in the membrane (and this means basically CD4 T helper cells) are directly killed by HIV. Cells like macrophages and dendritic cells, with relatively low concentrations of surface CD4, happily churn out large numbers of freshly made HIV without any apparent harm to the host cell. Why this should be so is unknown at present.

Because we can observe CD4 cells being killed by HIV outside the body, in a test tube (in vitro), it is clear that HIV is indeed directly cytotoxic for human CD4 T cells. But the situation inside the body is likely to be much more complex. There is a strain of mouse that mimics very closely the SCID condition that felled David the "Bubble Boy." These mice have no functional T or B cells of their own, and scientists have figured out a way to reconstitute them with human T and B cells. When these "SCID-hu" mice, reconstituted with human CD4 T cells, are infected with HIV, the CD4 T cells disappear, just like in human AIDS. But remarkably, the CD4 T cells disappear equally readily *whether or not the HIV strain is highly cytotoxic in vitro*. In fact, some strains that were not at all cytotoxic in in vitro assays were among the most effective in clearing CD4 T cells in the mouse. These results more than any others compel scientists to look for noncytotoxic mechanisms in the eventual development of AIDS in HIV-infected humans.

Inside the body, HIV almost certainly does kill CD4 T cells directly. But there is good evidence that HIV may also induce in its victims a form of immunological suicide. Remember that a major task of the immune system is to rid the body of virally infected cells. If the infected cells are themselves part of the immune system, the same rules apply. We have already seen what T cells do to macrophages and lung cells infected with intracellular bacteria like those causing tuberculosis. The T-cell-mediated killing of HIV-infected cells in the brain, as we saw earlier, is very likely responsible for the neurological deficits seen in many AIDS patients. So why wouldn't T cells do the same to each other? They do. The CD8 T cells that kill other T cells infected with HIV have in fact been demonstrated during the progression of AIDS. As in so many other situations of immunopathology, we realize that the immune system is simply following a predetermined program. Like some sort of robotic killing machines, T cells continue to lash out according to instructions, felling anything and everything in their path that is different. Even each other.

As noted previously, individuals infected with HIV are seropositive throughout the prolonged Stage II of the progression toward AIDS. This means quite simply that the immune system is fully cognizant of the presence of HIV in the lymph nodes, and is producing antibody and T cells that are at any given time specific for the strain of virus being produced. Yet here, once again, it seems very likely that a highly active immune response is actually responsible for the ultimate emergence of a form of HIV that can kill the host. Clearly the strains of HIV found in any given infected patient early in the response are not causing irreparable damage. There certainly is no wholesale destruction of CD4 T cells. But these relatively benign HIV variants are vigorously pursued and destroyed by the immune system, creating room for new variants to try out for the role of ultimate killer. The immune system does not make any distinction between meek and aggressive forms of HIV. It simply tracks down and kills any virally infected cell, wiping out whatever HIV variant might be inside. Eventually, a strain of HIV emerges that somehow manages to escape whatever immune surveillance mechanisms the immune system is throwing at it. As we have seen over and over in immunologically based diseases, it is a case of the blind leading the sighted. And once again, the sighted lose.

Preventing and Treating AIDS

Treating AIDS is a real nightmare for both the patient and the physician. The physician has to treat a wide range of problems simultaneously, and usually needs several specialists to assist him or her. Treatments that might ideally be used to treat two different conditions may antagonize one another or lead to levels of toxicity unacceptable in a human patient. The AIDS patient is assaulted by all of these treatments, which may involve considerable distress and disruption of normal body functions; the patient tries to believe they are all for the best but knows deep down that it may all be to no avail.

The principal question in the treatment of any disease, including AIDS, is: What is the goal of treatment? In the case of AIDS, is it to rid the body completely of all traces of HIV, or is it to simply make the presence of the virus in the body tolerable? Either of these would likely be acceptable to someone infected with HIV, but they represent two quite different challenges for treatment design. It may never be possible to rid the body completely of the virus. Remember, HIV integrates itself in a DNA form right into the patient's own DNA; it becomes a chemically indistinguishable part of the cell it infects. There is no known way to identify this DNA, or to remove it from the cell. If the DNA expresses itself, and directs the synthesis of foreign proteins the immune system can detect, then the immune system can at least potentially seek out the infected cell and destroy it. But infected cells remaining in the *dormant stage* are for all practical purposes invisible to the immune system. On the other hand, cells in the dormant stage are no threat to the host—unless of course the virus wakes up! So we begin to see some of the complexities involved in designing a treatment.

Once it was recognized that the central immunological problem in AIDS is the depletion of CD4 cells, numerous strategies to restore CD4 function were proposed, although never really tried. The futility of introducing a new source of CD4 T cells was apparent from the very beginning. There are many reservoirs of HIV in the body of an infected person. New CD4 cells, however they were introduced, would be useless; they would immediately become infected by residual virus. Then how about simply replacing CD4 T-cell function? As virtually all the known functions of CD4 T cells are mediated by the battery of lymphokines they produce and secrete, perhaps it might be possible simply to supply the body with external sources of these lymphokines, most of which are now commercially available. The problem is that these lymphokines are provided by CD4 cells to other cells in a tightly regulated fashion, exactly when they are needed, only in the place they are needed, and only in the amounts they are needed. Simply dumping them wholesale into the bloodstream would create immunological chaos. In fact, as AIDS progresses and the number of CD4 T cells diminishes, it appears that the residual CD4 cells are already in a highly activated state, pouring out their lymphokines without regard to time, place, or amount. This may be part of the problem. Increasing this chemical noise isn't going to help.

At present the major approach to treating AIDS that involves immunology is to create a vaccine against HIV. The usual strategy for vaccination is to use an attenuated form of the pathogen, with greatly reduced infectious potential, to induce "natural" immunity in the host. It is hoped the attenuated pathogen will stimulate the same vigorous type of immune response that the fully active pathogen does, without causing disease. However, the extremely high mortality rate of people infected with HIV has discouraged most vaccine designers from taking this approach. If even one HIV particle in a vaccine injection were incompletely inactivated, the results could be disastrous. More recently, through genetic engineering it has been possible to reproduce fragments of viruses for use as vaccines. Where successful, as in the case of hepatitis B, this method is preferred because there is no chance of the vaccine itself causing an infection.

Would a vaccine even be useful in dealing with AIDS? Clearly everyone infected with HIV does make antibodies to the virus, but the disease progresses anyway. Why should antibodies produced in response to vaccination be any more effective? This is an important question, and it has probably discouraged more than one drug company from climbing on the vaccine bandwagon. A major problem is deciding which HIV surface antigen to direct the vaccine against. Obviously those surface antigens that mutate rapidly are useless for vaccine development. Thus, the search has been for HIV surface antigens that are highly conserved *and* accessible to antibody. (Potential antigens buried *inside* the virus are useless, because antibodies made against them would not be able to bind to the intact virus circulating in the bloodstream.) The site on the gp120 protein that binds to CD4 would be one obvious candidate. This molecule cannot vary among different virus strains, or the ability to infect CD4 T cells would be lost. The surface proteins of HIV continue to be subjected to the most detailed biochemical analysis possible in search of candidate vaccine antigens.

The object of a vaccine administered prophylactically (before infection) would be to produce antibodies that would soak up HIV particles that make it into the body *before* they can infect a target cell. But this may be tricky. Antibodies make it easier for macrophages to pick up the virus and consume it. Viruses and other pathogens picked up this way are usually destroyed within the macrophage. But what if they weren't? There certainly are bacteria that manage to escape destruction inside macrophages. Remember tuberculosis? If even one HIV particle made it through the macrophage's digestive system, we have already seen that the macrophage is an ideal reservoir for HIV. What then? The possibility that a vaccine may actually enhance establishment of a virus has been observed in the past; the vaccines had to be withdrawn. Fortunately, the diseases involved were not as uniformly deadly as AIDS. Unfortunately, there is no way to predict in advance how a given vaccine will act.

At present, drug treatment for AIDS aimed at crippling the virus itself centers around a single drug, azidothymidine (AZT), also known by its trade name, Zidovudine. Although there are now more than a dozen drugs approved by the Food and Drug Administration (FDA) for treating various aspects of HIV infection, the only one that has made a significant impact on patient survival is AZT and related compounds. Originally developed as a potential anticancer drug, AZT was approved for treating AIDS patients in 1987. Since its introduction, AZT has effectively doubled the life span of persons diagnosed with AIDS—unfortunately, only from about one year to two. But that is a start.

The drug AZT is a slightly mutated form of one of the building blocks of DNA called *thymidine*. When AZT is incorporated into DNA in place of thymidine, all further synthesis of DNA stops. The advantage of AZT is that normal cells in the body cannot use AZT very well in place of thymidine. But viruses like HIV *can* use AZT. If AZT is present in a cell when HIV is trying to make DNA copies of its RNA genetic blueprint, AZT will be preferentially incorporated into the HIV DNA copies, and viral DNA synthesis is quickly halted.

Sounds simple, right? So why doesn't it work? If HIV cannot make DNA copies of itself, then HIV replication should stop dead in its tracks, and the infection should be history. In fact, as far as we can tell AZT *is* highly effective when first used. Most clinical symptoms of AIDS show some improvement, or at least stop getting worse. The downward slide in CD4 T-cell counts is arrested, at least temporarily, but after about a year or so, progress toward a more serious form of AIDS begins anew. The rate of degeneration once that happens is about the same as for someone receiving no treatment at all. What goes wrong?

The explanation almost certainly lies in the incredibly fast mutation rate of microbial pathogens in the face of strong negative selection pressure. The infection limps along on the one in a thousand, or one in a million viruses that somehow manage to survive in the presence of AZT. Through genetic mutation, strains of HIV eventually evolve that can use AZT *without* inhibiting DNA synthesis. Tests have shown that HIV strains present after a year of treatment with AZT are over a hundred times more resistant to AZT than were the strains present before treatment was started. Further drug treatment at that point is completely useless. Because of the nature of the attack mounted by AZT, the drug is powerless once HIV has successfully integrated itself into host cell DNA; it is harmless to already infected cells. Thus, AZT simply buys a little time; the eventual outcome is unchanged. The initial success with AZT led to its being given to AIDS patients at an earlier stage in the progess of the infection (the so-called Concorde trial), in the hope that a more profound effect could be obtained. Unfortunately, this hope has not been realized. After several years of follow-up, there appears to be little difference in the rate of progress toward full-blown AIDS between those receiving AZT early in the course of infection and those receiving no treatment at all.

There are other drugs currently in use, such as ddI (*Dideoxyinosine*) and ddC (*Dideoxycytosine*), that act in much the same way as AZT but have fewer side effects. AZT can cause headaches, nausea, and a severe form of anemia, all of which limit the doses that can be used. Only about half of all AIDS patients can tolerate AZT for more than a year. Both ddI and ddC are less toxic, but they too drive HIV to mutate drug-resistant forms. The current strategy is to give combinations of AZT and ddI or ddC, in the hope that it will be much harder for HIV to develop two simultaneous mutations conferring resistance to two different drugs. This may buy additional time, but again the eventual outcome for the AIDS patient will not likely change.

So drug treatment at present for AIDS is purely *palliative* (granting temporary relief from symptoms), and not at all curative. Currently, there are about a hundred drugs in various stages of testing that may have some impact on the progress of AIDS, but none of these are likely to provide a cure for AIDS. In desperation, some AIDS patients are turning to a wide range of implausible treatments. Who can blame them? The medical establishment can offer them little hope at present. In the case of both the body's own immune attack against the virus in its early stages and current drug treatments, the effect seems to be simply the generation of ever more resistant, ever more difficult to treat, substrains of HIV.

Not surprisingly, the effort to treat AIDS has also hastened the development of drugs that combat opportunistic infections. These are, as we have seen, the major cause of death in AIDS patients. For example, *P. carinii* pneumonia has responded well to aerosolized forms of the drug pentamidine, and CMV infections respond well to gancyclovir. Cryptococcus infections can be managed with amphotericin B. But in a truly bizarre and tragic twist of fate, most patients cannot tolerate most of these drugs at effective dosages at the same time they are being treated with AZT because of extreme bone marrow toxicity. Hence, both physician and patient are often left with the cruel choice of forgoing AZT treatment, which can definitely prolong life, or forgoing treatment of an infection that could well be fatal.

What Lies Ahead?

What is it going to take to get on top of this real-life "Andromeda strain"? Will we ever be able to cure people with AIDS? Will we ever be able to prevent people from being infected by HIV and developing AIDS in the first place? It must be admitted that at the present time there is no means to achieve either of these goals, nor are there any solutions based on present knowledge that immediately suggest themselves as likely prospects. Does that mean we are without hope? Not at all. Scientific breakthroughs—the sudden acquisition of new knowledge that either immediately solves a problem or makes it clear that a solution is possible and simply a matter of time—happen very frequently in biology and medicine. Most of these breakthroughs have little impact on the daily life of the average citizen (or even the average scientist, for that matter), and are rarely noted in the popular press. But as a part of the scientific process, breakthroughs are more common than people might think. They often come from the most unlikely sources—research on a fruit fly, perhaps, or through studies on how yeast cells reproduce.

What are some of the areas in which breakthroughs in the HIV-AIDS problem might occur? This gets us into an area where most scientists prefer not to go, but let us place a few possibilities on the table anyway.

A Solution to the Dormancy Question. As we saw earlier in this chapter, there is a period right after initial HIV infection when HIV has just inserted its DNA into the target cell genome to become a provirus, called the *dormancy period*. From that point until some unknown signal activates the provirus, HIV is competely dead within the cell. Absolutely nothing happens. What is the explanation for this? If we knew what was preventing the virus from being expressed during this period, or if we knew what element was still missing to trigger its activation, what a powerful tool that might be to manage an HIV infection! There is no guarantee, of course, that the activation process once initiated could be halted, but it is a question well worth pursuing. If we could just stop the virus from being expressed, all of the sequelae of AIDS would disappear, or maybe never happen. The T cells and other cells harboring HIV don't live forever. Eventually they die and are replaced by new cells of the same type. It is not inconceivable that, if activation could be prevented, we could eventually get rid of all the HIV-infected cells in the body, either by natural turnover or by drug eradication, thus achieving a true "sterilization" for this pathogen. What is needed is a clearer understanding of the life cycle of HIV, and how dormancy fits into it and is regulated. This is classical basic research, exactly the kind of tinkering around in the lab that scientists love to do just to see how something works.

A Vaccine. We previously mentioned some of the reservations about producing an AIDS vaccine, and the likely efficacy of such a vaccine were it produced. Despite these problems and reservations, some twenty different drug companies are vigorously pursuing the development and testing of over thirty vaccines. They are spending millions of dollars with no clear assurance of any return. Several of these vaccines are already in clinical trials, some involving hundreds of volunteers, testing potential toxicity of the vaccine material. These are all individuals already infected with HIV; none of the trials currently under way are aimed at uninfected individuals in an attempt to prevent HIV infection in the first place. That will come later. Although many of these vaccines appear promising in theory, it will be many years before we know if they are working. Most of the vaccine trials involve individuals in the early stages of infection. Because most of these subjects would not be expected to develop serious symptoms of AIDS for six to ten years, it will be some time before we know how effective the vaccines are in preventing disease onset.

At the present time, progression of vaccine trials into the large-scale, so-called Phase III trials that would actually begin testing vaccine efficacy on large numbers of uninfected but at-risk individuals have been put on hold by the National Institutes of Health. For such trials to be meaningful, numbers of volunteers on the order of six thousand to ten thousand would have to be recruited. Disagreement among scientists about the likely efficacy of even the most promising antibody-producing vaccines, coupled with doubts expressed by AIDS activist groups, has led to a delay of two to three years before such trials are likely to be initiated. But they almost certainly will be, either in the United States or possibly in some of the more stricken African or Asian populations.

Strategies are also being developed for *T-cell vaccines*. This is the approach favored by, among others, Dr. Jonas Salk, the developer of one of the polio vaccines. T-cell vaccines would try to induce CD8 T cells that more vigorously pursue and destroy HIV-infected cells, in an attempt to rid the body of actively expressed virus early in the infection. Of course, these are the very CD8 T cells that we said earlier may be responsible for the destruction of CD4 T cells that leads to AIDS. Wouldn't this be risky? Maybe not. If such CD8 T cells were present early enough in the infection, and in large enough numbers, they might be able to overwhelm the small number of HIV-infected CD4 T cells before the infection gets out of hand.

There is good reason to think such an approach might work. A few years ago immunologists took a close look at certain high-risk individuals who clearly had been exposed to HIV through sexual contact or intravenous drug use, yet who had not developed any signs of infection. The numbers of such individuals, though small, were of great interest to both scientists and physicians. Why weren't these individuals infected? Although they were seronegative—that is, showing no signs of having made antibodies to HIV—many of them did show very definite CD8 T-cell reactivity to HIV. Clearly they had been exposed to HIV, but got rid of it before the infection even got to the antibody-inducing stage. Were the HIV-sensitive CD8 cells in these individuals responsible for this evasion? Some vaccine designers—including Jonas Salk—are betting this was indeed the case. At this stage, no one is ready to dismiss out of hand *any* approach to vaccine development. Wouldn't it be terrific if we could immunize for antibody formation *and* the induction of CD8 T cells, simultaneously? Why not?

Drugs. Drug development is in many ways in the same state of darkness and uncertainty as vaccine development, but that does not mean that research isn't going ahead full-steam anyway. The complete life cycle of HIV is known in possibly greater detail than any other virus on the face of the earth. As we saw earlier, it attaches to a cell; enters it, undresses; converts its RNA to DNA; inserts its DNA into the host DNA; uses this DNA to make new viral RNA; dresses this RNA in a protein coat; and sends a thousandfold excess of mature viral particles out of the cell. Each of these steps is a potential point of attack. AZT is aimed at only one of them, conversion of proviral DNA to RNA. Scientists are busy at this very moment trying to find drugs that zero in on some of the other steps in the HIV life cycle. The secret is finding drugs that confound the virus without being unduly toxic to healthy human cells. If even one additional drug is found, attacking a point in the HIV life cycle different from the one targeted by AZT, the combination of two such drugs administered simultaneously could provide a powerful defense against HIV infection.

The approach just described is what scientists like to call rational drug design. By studying the most intimate details of HIV's life cycle, scientists try to develop a specific drug designed to exploit an inherent weakness in the way HIV goes about reproducing itself. That is how AZT and related compounds were found. Another approach to new drug development is the so-called shotgun approach. Forget understanding how the virus works. Just throw everything imaginable at it and see what works. If you find something, let the pointy heads figure out *how* it works. You just want something that stops this virus dead in its tracks. Even if you just slow down its replication rate, you could decrease the rate at which new mutant variants are formed and give the immune system a better chance of doing its job. Frankly, that is how a great many drugs currently used in the clinic were discovered. They may seem like triumphs of the intellect; they may very well just have been a shot in the dark on the part of someone who hadn't the foggiest idea of what he or she was doing.

Both these approaches are perfectly valid, and the one that a particular scientist or group of scientists takes is largely a matter of personality and style. Either could work. Drug companies are spending millions of dollars in both directions. By the way, the "SCID-hu" mouse described earlier, in which a SCID mouse is reconstituted with human CD4 T cells and then infected with HIV, is proving to be a valuable tool in testing candidate HIV drugs, whether rationally or irrationally designed. The ability of drugs to interfere with HIV replication, or HIV cytotoxicity against human CD4 T cells in these mice, is looking more and more like a reliable predictor of what these same drugs might do in humans.

At the present time we have absolutely no idea when a breakthrough might occur in producing a drug that will work against HIV. It could be next week, next year, or never. Companies investing in this type of research may never realize a penny on their investment. That is one reason why aspirins, which cost less than a tenth of a cent to produce, end up costing a nickel apiece at the drug store.

Gene Therapy. The major problem with both vaccines and antiviral drugs is that as the targeted pathogen undergoes perfectly normal replication, it may produce an occasional mutant progeny that is resistant to treatment but still pathogenic. Under the selective pressure of the treatment itself, a single mutant pathogen—maybe one in a billion or one in a trillion or more of the total population—can grow to dominate the entire pathogen population in a short time, replacing the susceptible forms of the pathogen and rendering the drug or vaccine useless. But what if the very cells the pathogen attacks could be equipped with a means of destroying the pathogen the instant it began to replicate? It would never have a chance to produce mutant progeny that could escape the treatment. This is the basis of the *gene therapy* approach to fighting AIDS.

The strategy that would be used is similar to that described earlier for treating primary immune deficiencies. Like primary immune deficencies, the cells affected in AIDS—T cells and macrophages—are ultimately derived from stem cells in the bone marrow. In the latest treatments for disorders like SCID (which, you may remember, is very much like AIDS in its consequences for the patient), bone marrow cells were removed, and a "good" copy of a defective gene in the patient's marrow cells was introduced—a sort of molecular repair job. In the form of gene therapy commonly envisioned for AIDS, a different kind of gene would be slipped into stem cells—a gene that would stop replication of the virus dead in its tracks. If the virus cannot replicate, it cannot produce mutants. This approach is sometimes called "intracellular immunization" because it brings the level of protection right into the infected cell itself.

A number of "immunizing genes" have been envisioned. One fairly simple possibility would be to introduce a gene whose RNA message (the strand of RNA copied from a gene and taken out into the cell's cytoplasm to direct the synthesis of a needed protein) is an exact mirror image of one of the RNA messages the virus makes in order to reproduce itself. The presence of this mirror-image RNA would have no consequences whatsoever for a normal, healthy (uninfected) cell. It would not be used for anything, and in time would disappear, to be replaced by more mirror-image copies that are equally harmless. But if the cell became infected by HIV, the mirror-image RNA would suddenly become very valuable. Messenger RNA can carry out its function *only* as a single strand. Mirror images of RNA will bind tightly together, forming RNA double strands. So the crucial viral RNA and its mirror-image RNA, inherited from the treated bone marrow, would bump into each other in the cytoplasm and instantly form double strands. These double strands are recognized by the cell as "mistakes" and destroyed. No matter how many of the "real" RNA messages the virus tried to make, they would always be neutralized by the inserted mirror-image messages and immediately eliminated, with absolutely no harm to the cell.

A number of variations on this basic scenario are being tried in the laboratory in other, less deadly viral systems to see if they work. So far the results are highly encouraging. The beauty of this approach is that individuals already infected with the virus could be treated as effectively as uninfected persons. In an infected individual, when the virus escaped from the dormant phase it would begin killing off CD4 cells that had been derived from untreated marrow. Nothing could be done to save those cells. But as those cells were replaced with new CD4 cells from marrow stem cells carrying the rescue gene, the virus would gradually be deprived of a place to replicate and gradually disappear from the system.

The use of molecular biological approaches in treating a wide range of human diseases, but particularly those affecting blood cells, is now upon us. Gene therapy approaches to treating AIDS is being studied intensively in virtually every university medical center in the United States. Serious clinical trials should get under way in the near future. It is entirely possible that halting the AIDS virus may be one of the most spectacular successes of gene therapy to date. Research on this important topic is proceeding at a furious pace.

"The Worst Possible Nightmare"

But what if there is no breakthrough? It is estimated that at the present rate of increase 100 million—2 percent—of the world's inhabitants could be infected with HIV by the year 2000. What if we are left to our natural biological selves to deal with this modern plague, with no help from science or medicine? Is there a chance we will ultimately develop a natural resistance to HIV not dependent on the immune system, or on external drugs or vaccines?

Theoretically, if HIV began seriously decimating the human population, this *could* happen. But the cost could be very high indeed. At present, HIV transmission from one individual to another can take place only under highly restrictive conditions, mostly based on the direct mixing of bodily fluids. But what if a strain of HIV suddenly emerged that could be transmitted between individuals through the air? An HIV carrier who sneezed on an elevator could infect the next dozen or so people getting on. In the course of a common cold, with all of the attendant coughing and sneezing, he or she might infect a hundred or a thousand people. That is exactly how colds themselves are spread. Given the long period of time before the individuals infected would know they are HIV-positive, transmission could move outward to infect thousands more.

This scenario is the worst possible nightmare with respect to AIDS, but unhappily it is not entirely beyond the realm of the possible. Under such conditions, individuals with, say, spontaneous mutations in their CD4 molecules that deprived gp120 of a binding site could come to have a selective advantage. The same would be true of any other human mutation that interfered with HIV reproduction. Perhaps this is not an idle speculation. Evolutionists have focused in recent years on something called *punctuated equilibrium*. The greatest evolutionary changes seem not to occur slowly, through the accumulation of minor mutations over time, but very rapidly, usually in response to some catastrophic environmental alteration. The extremely rapid replacement of dinosaurs by later forms of vertebrates, for example, appears to have occurred in the aftermath of a meteor reaching the earth's surface some sixty-five million year ago, at the end of the Cretaceous period. In evolutionary terms, this all happened in the blink of an eye. It takes little imagination to picture the consequences wreaked by such enormous devastation in the biosphere. More than half the animal life forms existing on earth at the time—including most large land animals—are thought to have disappeared. Life-forms with characteristics that gave them even a small survival advantage at all came to dominance in very short order in this new world. These changes took place over such a short period in geological time that there is virtually no fossil record of the enormous range of early and intermediate mammalian life-forms that emerged.

Similarly, if the human population were reduced to very small numbers by HIV, it is entirely possible that the earth could see another example of punctuated equilibrium. Humans could either be extinguished altogether from the earth, or a few individuals with chance mutations somehow protecting them from infection by HIV could reproduce and ultimately give rise to a new strain of *Homo sapiens*. Whatever genetic changes that allowed them to evade infection by HIV would dominate the new strain completely. Even if HIV subsequently disappeared from the face of the earth, these changes would likely remain indefinitely as a sort of genetic "fossil record" of the HIV experience. So, the answer is yes, we *could* develop a natural resistance to HIV not dependent on the immune system, but only under conditions so catastrophic that human life on this planet, as we know it now, would be altered beyond recognition.

In our society, AIDS also highlights another aspect of evolution that is often forgotten in the talk about genes, breeding, and natural selection, and that is *cultural evolution*. Although we clearly evolved according to the same laws as any other animal group, human beings (perhaps uniquely among all living things on this earth) are probably no longer subject to natural selection and evolution as originally described by Charles Darwin, at least in the absence of catastrophic events that produce punctuated equilibrium. Natural selection is the process whereby those individuals best fit to compete for limited resources in the environment gain a reproductive advantage. When they die, they leave behind more offspring carrying their genes than do others of their species. Human beings have gained a sufficient level of control of environmental resources in most parts of the world that competition for these resources is no longer a factor in reproductive behavior. The dominant factors controlling human reproductive advantage are now cultural; they have relatively little to do with an individual's genetic makeup. As far as we know, they also have nothing to do with survival of the species as a whole. Unlike biological evolution and natural selection, which are incredibly slow for species with long life spans, cultural evolution, and the selection and stabilization of the changes it produces, is extremely rapid. To the extent that survival and reproduction are affected by *cultural* evolution, we may indeed be able to outwit rapidly evolving pathogens.

Nowhere could this be clearer than in the case of AIDS. We have already produced drugs that slow the growth rate of HIV. Tomorrow, or next week, or next year, humans may produce a vaccine or a drug or a gene-therapy strategy that absolutely stops HIV dead in its tracks. That has nothing to do with human *biological* evolution, or manipulation of a naturally evolved immune system; it is simply the application of tools developed through cultural evolution. Moreover, impressive inroads in preventing AIDS through behavioral modification have been made among at least some segments of the population. These gains have been achieved not by manipulation of the virus or the host at a biological level, but again at a cultural level, through information and education, and self-imposed behavioral changes. In the end, if the built-in defensive genes we have cannot save us, and if we cannot produce new ones at a competitive rate, we may find that what saves us is the application of our own intelligence and accumulated cultural wisdom. There is precious little a virus— even one as deadly as HIV—can do about that.

SEVEN

Organ Transplantation: Exploring the Boundary Between Technology and Ethics

Late in the summer of 1954, Richard Herrick was referred by his doctor to the Peter Bent Brigham Hospital in Boston, Massachusetts. Richard was twenty-four years old and had been suffering for some time from high blood pressure and puffiness around the face and eyes. His doctor suspected a kidney problem, and the Brigham is where people went if their kidneys were malfunctioning. The medical staff at the Brigham ran a battery of tests that initially might have indicated any number of problems. But they noticed that in addition to high blood pressure Richard had a bit more protein than normal in his urine, as well as traces of blood. Together with other findings, this confirmed the diagnosis of a kidney dysfunction. Richard was transfused with several units of blood, which improved his condition considerably, and he was sent home. Only time would tell how serious the problem with his kidneys was.

Five months later, Richard Herrick was back, and this time it was clear he had a very serious problem indeed. His blood pressure was now dangerously high, and he was beginning to experience problems with his vision, a not uncommon by-product of high blood pressure. Protein levels in his urine were double what they had been before, and he was showing signs of congestive heart failure. Several days after this second admission Richard began to exhibit bizarre behavioral changes; he occasionally became drowsy and disoriented; at other times he was irritable or even aggressive toward the staff. He went into convulsions several times. It was a set of symptoms the doctors at the Brigham were all too familiar with, and about which they knew they could do precious little. Their young patient was experiencing the beginning stages of massive and terminal kidney failure.

Dr. John P. Merrill took a special interest in this particular patient. Merrill had been working with a medical equipment company on the refinement of an "artificial kidney," what we would today call a renal dialysis machine. This machine, first developed in Holland during World War II, was showing great promise in being able to substitute for one of the most vital kidney functions—removing from the blood toxic substances that could cause precisely the symptoms this young man was experiencing. In fact, on his second visit to the hospital, Richard was treated with one of the artificial kidneys and, as the doctors expected, showed great improvement.

But another chance to demonstrate the usefulness of his new machine was not what attracted Merrill to this case. Merrill knew that the kidney machine could never be more than a stopgap measure, able to keep a patient alive for a period of time but never able to offer a cure. What he was really interested in was the possibility of kidney transplantation. He had recently completed a series of nine kidney transplants, taking healthy kidneys immediately after death from patients who died of causes unrelated to their kidneys, and transplanting them into patients with terminal kidney failure. In several cases, the transplanted kidney had seemed to take hold for awhile, bringing almost immediate improvement in the recipient's condition. But in a fairly short time all nine transplants had failed, and the recipients all ultimately died of terminal kidney failure. This was incredibly frustrating, because Merrill's team was considered to be one of the most skilled in the world at this procedure.

Like other experts in his field, Merrill was convinced the transplants were failing not because of problems with the surgery, or because an organ from one person simply could not function in another, but because the transplanted organ was being attacked and rejected by the recipient's immune system. Among the various lines of evidence in support of this notion, he had been particularly struck by the experiments of Ray Owen, who had shown that twin cows that shared a single placenta during fetal life could exchange grafts as adults. Merrill had argued for some time that human identical twins should be able to exchange organs and tissues without any fear of immunological rejection. And that was what interested him about this young man. According to the doctor who had referred Richard to the Brigham for treatment, Richard had an identical twin. After reassurances that he could survive with a single kidney, Richard's twin agreed to give the new procedure a try. Merrill and the Herrick boys were about to make medical history.

As a preliminary test of his hypothesis, Merrill's team carried out an exchange of skin grafts between Richard and his twin brother. After a rather anxious month in which his doctors had to struggle to keep Richard alive, it was confirmed by microscopic examination that he had completely accepted his brother's skin. Without waiting any further, the two brothers were prepped and wheeled into adjacent operating rooms. The left kidney from the healthy twin was removed and taken in a stainless steel pan to the surgeons waiting in the adjoining operating room. While the first twin was being closed, the surgeons opening Richard saw a sight usually only seen at autopsy—two shriveled, shrunken kidneys wasted away to a tenth their normal size. Although the healthy twin's kidney had grown pale and cold during the eighty-odd minutes between operations, as soon as it was connected to Richard's circulatory system it swelled ever so slightly and turned pink and warm to the touch. After the doctors checked meticulously for leakage, this young man, who only days before had been within a stone's throw of death, was carefully sewn back together. Recovery from the surgery was uneventful for both brothers, and the transplanted kidney began to function beautifully in its new surroundings. All of Richard's previous symptoms disappeared in a matter of days. He was discharged after two weeks, and over the course of the next few months regained his former physical vigor, as well as twenty-five pounds of lost weight. The donor brother's remaining kidney underwent a gradual enlargement as it took on the sole task of cleaning out his blood, but he suffered no ill effect whatsoever. Both brothers lived for many years.

Thus began the age of human organ transplantation. Of all the miracles wrought by modern medicine, none has moved us quite the way organ transplantation has. That an organ can be severed of all its connections with one human being, implanted into another, and recover the full function it needs to sustain life in the recipient was and remains simply awe-inspiring. When, as in the case of bone marrow or a kidney, both the donor and the recipient may be alive and well after the transplant has been accomplished, a bond is established between them that is unique in the human experience. On the other hand, to see a transplanted heart still beating and sustaining life in a human being a quarter century after its original owner has returned to the elements he or she came from puts us in very close touch with some of the deepest mysteries of life, and it stretches our conception of the meaning of mortality and immortality. How did we come to be able to do such a miraculous thing?

Our fascination with the possibility of using transplantation to restore broken or worn-out body parts seems to have been around for a very long time. Ancient medical texts describe attempts to replace at least the external parts of the body. As early as several hundred years B.C.E., Hindu surgeons described a technique for reconstructing noses from tissue obtained elsewhere in the body. This was necessitated by a fairly common punishment for a number of crimes in ancient India: cutting off the nose. One of the earliest accounts of transplantation in Western culture, although certainly apocryphal, suggests that such experiments may at least have been thought about. Cosmas and Damian were two third-century Roman physician-brothers who had the strange practice of not charging for their services. They were eventually beheaded for their erratic behavior, which also included conversion to Christianity. They are alleged to have returned some two hundred years after their execution to a church in Rome dedicated to their martyrdom, where the caretaker of the church had apparently developed gangrene in one of his legs. According to legend the brothers removed the bad leg and transplanted a good one from a recently deceased Moor. We are left to believe that this worked, and that the caretaker went about ever after with one black leg and one white leg.

Gasparo Tagliacozzi, a sixteenth-century Italian surgeon, described a method for using tissue taken from the arm to rebuild a nose. The arm is brought up and fixed into position next to the nose; an appropriately shaped slice of muscle with its overlying skin is gradually carved away from the arm and allowed to implant on the face. After the tissue has been finally severed from the arm and is settled into its new location, the arm is lowered and allowed to heal. With only minor variations, this technique is still used today, and is called the *Tagliacozzi flap* procedure. The use of tissues taken from one part of the body to reconstruct or repair another part of the same body, called *autografting*, is strictly a surgical problem. With proper technique, any part of the body should be transplantable to any other part of the body; whatever barriers may exist are clearly not immunological in nature. The exchange of body parts between two genetically different individuals (*allografting*) is, the Miracle of the Black Leg excepted, quite another matter.

Of all the surgical techniques associated with organ transplantation, the most critical is *vascular anastomosis*, or the suturing together of blood vessels between the donor tissue and the recipient's blood system. Every organ in the body is intricately connected with the body's circulatory system. Each organ is served by arteries, which bring fresh blood to it, and by veins, which take used blood away from it. If this circulation is interrupted for more than a few minutes, the organ will suffer irreversible damage and die. It is very easy to disconnect the arteries and veins when removing an organ; a surgeon can just snip them with a scissors or slice them with a scalpel. Damage to the organ owing to removal from its oxygen supply can be minimized by cooling. But reconnecting an organ to the recipient's circulatory system is very demanding, almost an art as much as a science. The intricate methods for achieving this were not developed until the turn of the twentieth century, by a surgeon named Alexis Carrel. The techniques he developed at that time, specifically for the purpose of transplanting organs, were used by John Merrill almost exactly as he first described them. Carrel thus made two new fields possible: vascular surgery and, indirectly, organ transplantation.

Carrel and other surgeons in fact spent the next thirty years exploring the surgical aspects of organ transplantation in animals, and they made great progress. Moving organs from one place to another in the same individual, and getting them to function, proved with practice to be fairly easy. But in terms of their ultimate objective—transplanting organs from one animal to an *unrelated* animal and achieving long-term survival and function of the organ—they, no less than Merrill after them, were completely without success. What Merrill's very important experiment with human identical twins showed was that the surgical skills necessary to accomplish successful organ transplantation were well in hand, indeed, probably already had been in hand in Carrel's day. What remained was to remove, or at least to manage, the immunological barriers.

The Immunological Basis of Organ Transplantation

It seems intuitively obvious that human beings are all very different from each other, and that the immune system could possibly spot these differences and respond to them. But what exactly are the differences between people that the immune system responds to? These differences are clearly absent in identical twins, and present in everyone else. But are all differences the same? Might some people be closer in terms of these differences than others? And if so, is it easier to exchange grafts between them?

One of the most important steps toward understanding the immunological basis of organ transplant rejection was the gradual unraveling of a concept that came to be referred to as *histocompatibility*. *Histos* is a Greek word referring to a weaving, or a web, and immunologists co-opted this term in combinatorial form to refer to the compatibility of living tissues or organs from people or animals who are genetically different. Our understanding of histocompatibility stems from an interesting intersection between the fields of organ transplantation and cancer research. At the beginning of this century, researchers wanted very much to be able to pass tumors from one animal to another in order to study the process of tumor growth and development. The animal most often used in such studies was the mouse, which is small, relatively inexpensive, and easy to maintain in the laboratory. The problem with studying tumors in mice (or any other animal) is that although tumors may start out small, they keep on growing and eventually kill the animal carrying them. So as the tumor got larger, and the poor mouse carrying it might seem to be nearing the end, researchers would try to pass a small piece of the tumor to another mouse to keep the study going. Usually this would fail. The tumor would seem to grow for a day or two or three, and then shrink and disappear. But on rare occasions the tumor would "take" and could even survive passages through several consecutive mice. There was great speculation about the reason for this occasional success. The failure to "take" was assumed by many scientists to be due to some special property of tumors. But no one could discover what this property was, or why it worked in some cases and not in others.

At some point an alert lab worker apparently noticed that the more closely related two mice were genetically, the more likely it was that a tumor could be passed successfully between them. In a study carried out in Germany in the early 1900s, it was observed that a tumor arising in wild mice captured in a particular house could be passed with a substantial number of positive takes to other mice captured in the same house; with less success to mice captured in nearby houses; and not at all to mice captured in distant neighborhoods. The explanation of this lies in the sociobiology of mice. Mice living in any given household tend to be closely related, forming what is known as a *deme*. This is largely due to the murine equivalent of incest, which results in a substantial degree of genetic homogeneity. Occasionally, disgruntled males may leave one deme and bully their way into a neighboring house, establishing a distinct but genetically related deme just next door.

Around the same time, it had become fashionable to keep so-called fancy mice as pets for children. These were mice specifically bred, using techniques known to farmers for centuries, to bring out some property thought to be cute, or at least commercially profitable. This sort of inbreeding produced albino strains with white fur and pink eyes, for example—a real oddity at the time, although fairly commonplace now. Selective breeding also produced the famous "Japanese waltzing mouse," a poor creature with an inner ear defect that led it to stagger ("waltz") in circles in its cage. (This apparently was an example of cute.) But like mice living together in the same house, these partially inbred strains of mice showed a high degree of acceptance of each other's tumors.

And so one day, as happens from time to time in immunology, a light went on. The more closely related two individuals are, the more likely it is that they will be able to exchange tumors—or, for that matter, *any* tissue. And this led ultimately to the discovery of *histocompatibility antigens*. This is perhaps one of the most important discoveries in all of immunology, from both a practical and a theoretical point of view. We now know that histocompatibility antigens are special proteins found on the surface of each cell in the body. Every cell in the body of the same individual (whether mouse or human) will have exactly the same histocompatibility ("tissue compatibility") antigens on its surface, marking those cells as belonging to that individual. However, two different individuals (unless they are identical twins) will have different histocompatibility antigens.

In humans, the likelihood that two randomly selected individuals could have the same set of histocompatibility antigens (called *HLA antigens* in humans) is less than one in twenty million. These are thus truly "markers of individuality" and are the main reason human beings cannot exchange tissue grafts: The immune system is exquisitely sensitive to differences in histocompatibility antigens and will mount a vigorous and effective rejection response against any HLA antigens that are not self. Thus the inability to pass tumors from one animal to another turns out to be simply a variation of the general theme that tissue grafts cannot be passed between individuals. In both cases, it is the difference in histocompatibility antigens between donor and recipient that triggers rejection.

Although the odds of two randomly selected individuals being completely HLA identical are extremely low, it still helps to try to match them up as best we can. This is done by a process called *tissue typing*, in which the HLA antigens of prospective donors and recipients are identified, and the best possible match is made. There is a reasonably good correlation between the degree of HLA matching and success of the transplant. Especially when the donor and recipient are unrelated, every effort is made to achieve the closest possible HLA match between them.

The basis for graft rejection between nonidentical individuals was hotly debated throughout the first half of the twentieth century. As early as 1912 there were suggestions that it might be immunological in nature, but this was not immediately obvious to many people. Most early transplant experiments in animals involved the exchange of skin grafts, which are technically easy to perform. These studies demonstrated that graft-specific antibodies, although indeed produced during skin transplant rejection, had little or no effect on graft survival. As antibodies were at the time the only known immune effector mechanism, it was quite reasobable to conclude from these results that skin graft rejection could not be immunological in nature.

The fact that graft rejection is indeed immunological in nature was finally demonstrated to everyone's satisfaction by the British physician-scientist Peter Medawar during World War II. Medawar (later Sir Peter) was working at a burn hospital in London, treating civilians injured in bombing raids in England, as well as British soldiers and airmen returned from more distant fronts for advanced care. It had been known for some time that the most effective treatment for severe burns is to get the burned area covered as quickly and as completely as possible with fresh skin to prevent infection and loss of body fluids. Skin taken from another part of the patient's own body was known to be the best solution, but this was not always possible. Using skin from other donors could sometimes offer temporary relief, but the transplanted skin would always be rejected in the end. Nevertheless, foreign skin could sometimes last long enough to allow the scarring process in the patient's own underlying tissues to get under way.

In particularly bad cases, it would sometimes be necessary to apply a second transplant of skin to keep patients alive until their own healing processes could take over. Medawar noticed that a second application of skin taken from the same donor, after the first graft had been rejected, would last only a few days, whereas the first graft may have lasted up to two weeks. However, a skin graft from a completely *different* donor applied to a previously grafted patient would again last up to two weeks. So it had become common practice never to use skin from the same source twice on the same patient. Upon reflection, Medawar concluded that the skin graft recipient must have been mounting an immune reaction to the original transplant of skin. If skin from the same donor was transplanted a second time, then immunological memory came into play, and the graft was vigorously and rapidly rejected in the same way as a secondary infection with any standard pathogen would be. Medawar followed up his clinical observations, published in 1943, with a series of incisive skin-grafting experiments in rabbits that convinced everyone working in the field that graft rejection was indeed immunological in nature. The combination of his clinical and experimental studies on transplantation and tolerance resulted in a Nobel Prize (shared with Sir Macfarlane Burnet) in 1960.

Several years after Medawar's experiments, it was finally shown that skin transplant immunity was caused by white blood cells rather than antibodies, thus providing a rational immunological basis for transplant rejection. We now know that skin graft rejection is caused almost exclusively by T cells that recognize foreign histocompatibility antigens on the incoming graft cells. These T cells belong to a different subset than the ones we have seen previously. The cells that help B cells and macrophages, and that are attacked by HIV, are CD4 helper cells. The T cells that cause graft rejection are CD8 "killer" T cells, also called *cytotoxic T lymphocytes*, or *CTLs*. The result of CTL attack is swift and violent: The graft drips away, withered and dried, less than two weeks after transplantation.

From Pastime to Prime Time: The Advent of Immunosuppressive Drugs

By the end of the 1960s, physicians and scientists could add organ transplantation to the list of clinical situations in which the immune system was part of the problem, and not part of the solution. Organ transplants that could demonstrably save a patient's life are thrown out of the body as rabidly and as rapidly as any disease-bearing pathogen. True, the immune system is not doing anything wrong. Evolution had never prepared it to make decisions about spare body parts that are not part of self. The question became, would it be possible to somehow selectively disconnect the immune system with respect to a newly transplanted organ without, at the same time, crippling it with respect to its ability to fight infections?

The very first transplants in humans—the kidney transplants using identical twins as donor and recipient—were carried out without the need for suppression of the immune response of the recipient. Because both recipient and donor always had exactly the same histocompatibility (HLA) antigens, there was nothing to provoke an immune response. On the other hand, as every transplanter up through Dr. Merrill had found out the hard way, all transplants attempted with other than an identical twin donor—even if donor and recipient were closely related—involved HLA differences, and they failed because of immunological rejection. It seemed at first that transplantation might be limited to that small handful of cases in which an identical twin donor was available. Only a dozen or so such transplants had been carried out in the United States in the years immediately following 1954. It began to look as though transplantation would simply take its place on the shelf of medical and immunological oddities, of little use to society at large.

A few attempts at immunosuppression were made in those earliest days of transplantation. The only known way to suppress the immune system at the time was with radiation. High-energy radiation from sources like radioactive isotopes or X-ray generators was known to inhibit immune function, and in fact several transplants were attempted in the late 1950s and early 1960s using whole-body X-irradiation to prevent rejection. The level of radiation that had to be used to obtain an effect, however, was simply too toxic, especially toward bone marrow, to be tolerated, and this approach was soon abandoned.

And then one of those completely unforeseen breakthroughs occurred that virtually revolutionized organ transplantation overnight. Like the discovery of histocompatibility, it too was tied to cancer research. The new field of *cancer chemotherapy* had begun in the early 1950s with a deliberate attempt to synthesize drugs that would interfere with known metabolic pathways crucial to cancer cells in the hope of selectively halting their growth without affecting normal cells. For example, cancer cells divide very rapidly, and thus they synthesize DNA on average much more frequently than do normal cells. In the early 1950s, chemists began to synthesize drugs that might interfere selectively with DNA synthesis in cancer cells. One such drug was 6-mercaptopurine (6-MP). Like AZT, the drug used to treat AIDS patients, 6-MP is an analog of one of the building blocks of DNA, and it too can deregulate the normal synthesis of DNA. The development of 6-MP was a result of the process of *rational drug design* discussed earlier in connection with AIDS.

It was hoped that cancer cells, because of their high rate of DNA synthesis, might prove to be especially sensitive to 6-MP. Although 6-MP did prove to be modestly successful in that regard, its most important biological effect would be found to lie elsewhere. In 1959 it was reported that, unexpectedly, 6-MP could also profoundly inhibit the ability of animals to clear foreign proteins that had been injected into their systems. It was rightly suspected that this was due to an impairment of antibody synthesis. This was the first documented instance of chemical suppression of the immune response. The possibility that human beings could reach inside the body and manipulate the immune response with drugs opened the door on an entirely new era in immunology.

Tremendous excitement surged through both the medical and scientific communities as the implications of these findings for organ transplantation became apparent. Within a year 6-MP was used successfully in an attempt to prevent rejection of kidneys transplanted between unrelated dogs. The results were so impressive that barely a year later 6-MP was brought to the clinic for its first use in human transplantation. Although the first patient treated with 6-MP, a twenty-two-year-old male with end-stage renal disease who received a kidney from an unrelated cadaveric donor, lived only twenty-seven days, a medical record was set. Moreover, the patient died of a heart attack, not kidney failure, and his transplanted kidney showed no sign of immunological rejection at autopsy. Within a very short time, patients were surviving for several months, and then several years, as medical personnel became more skilled in administering 6-MP and managing its side effects.

These results, achieved in a remarkably short time, truly ushered in the modern era of organ transplantation. As we said earlier, organ transplantation was first made possible technically by the development of vascular surgery. But it was rescued from being a mere medical curiosity limited to identical twins by the advent of chemical immunosuppression.

Immediately after the introduction of 6-MP to the clinic, the number of centers venturing into this new area of medical razzle-dazzle blossomed overnight. Chemists were set to work to make derivatives of 6-MP that would be less toxic. One such derivative, azathioprine (trade name Imuran), became the standard of the transplant clinic for twenty years. Pharmaceutical companies everywhere began screening a wide range of drugs in animals to see if other acceptable immunosuppressants might be out there. Among the more effective were various corticosteroids that, when used in combination with Imuran, gave quite impressive results.

But this medical miracle would not be without its costs. The level of immunosuppression necessary to make transplantation successful can have serious side effects, and in the early days of transplantation these were often severe. The drugs used are almost all deadly poisons, originally developed in many cases to kill tumor cells. They are introduced into the body with the aim of selectively suppressing cells of the immune system involved in graft rejection, but there is absolutely no way to limit their effects *just* to cells of the immune system. Thus one notable limitation to the use of these drugs is the serious damage, unrelated to immunosuppression, they may do to any of a number of organs or tissues in the body, including the transplant itself. The bone marrow is especially sensitive.

The second problem with these drugs, and perhaps the more profound one, is that although the intent may be just to suppress those immune cells involved in rejection of the transplanted organ, they in fact suppress the immune system as a whole. The result, not surprisingly, is a secondary or acquired immune deficiency condition not unlike that seen in AIDS. It is characterized by infections with a wide range of external and opportunistic pathogens, and by abnormally high rates of cancer. The opportunistic pathogens causing problems in transplant patients are basically the same as those seen in AIDS: the fungi C. *albicans* and P. *carinii*, and various viral infections. For many years P. *carinii* pneumonia was the most common cause of death in transplant recipients and is still a major problem. A disturbingly high proportion of patients died from infection with their transplanted organ looking robust and healthy at autopsy.

In the early years of transplantation the major cancers seen were cancers of white blood cells. As these were the cells targeted by the immunosuppressive drugs used, it was thought that the cancers seen might be a direct effect of the drugs on the white cells, rather than a result of immunosuppression per se. But as patients started to live longer with their transplants, a much wider range of cancers began to be seen, suggesting that the immune suppression needed to prevent transplant rejection was indeed allowing cancers normally controlled by the immune system to break free and cause disease. Ironically, one of the cancers now seen most commonly in long-term transplant survivors is Kaposi's sarcoma.

Thus, by the late 1960s, transplantation across genetic differences was a reality, but reality with a stiff price. In the early 1970s, that price would go down dramatically. A team of scientists at Sandoz Laboratories, a Swiss pharmaceutical company, had been screening various soil funguses in search of drugs that could be used to treat fungal infections in humans. This is the process of "shotgunning" referred to earlier in connection with the search for new AIDS drugs. Drug companies are constantly scrutinizing nature's own pharmacy, looking for new medicines, in addition to using the kind of rational drug design that led to the development of 6-MP. The Sandoz scientists were working on the idea that one strain of fungus might produce a substance—an antibiotic—that it used to kill off other strains of fungus competing for the same environmental niche. They were not having much luck. Several compounds did seem to have some antifungal activity, but these did not look promising clinically because they were ineffective against those funguses that are serious pathogens for humans.

One of these compounds, which eventually came to be known as *Cyclosporin* A (CsA), had been isolated from a fungus growing in the soil in southern Norway. It seemed interesting because it had very low toxicity; it could also be used in animals at quite high concentrations without apparent side effects. One of the

Sandoz team members, Jean Borel, decided to carry out a wider examination of the pharmacological properties of CsA in other situations of potential clinical interest. What Borel found surprised him and put broad smiles on the faces of the directors of Sandoz that would last for years. Cyclosporin A turned out to be an incredibly potent immunosuppressant, equal to anything known at the time. But more importantly from a clinical point of view, it had a profound inhibitory effect on organ transplant rejection, with far fewer side effects and much less toxicity than the drugs currently in use.

When CsA was brought to clinical trial in 1983, the results were beyond Borel's wildest expectations. Prior to 1983, some 50 percent of kidneys transplanted from cadaver donors failed after one year. Almost immediately after the introduction of CsA, this number fell to 15 percent! The impact on heart transplantation was equally remarkable: Not only did the success rate nearly double, but the average hospitalization time fell from seventy to forty days, greatly easing the financial burden on the overall health care system.

Unlike 6-MP and Imuran, which allow transplants to survive by suppressing essentially the entire immune system, CsA acts specifically to block the activation of T cells. If it is present during the period when a T cell is encountering a particular antigen for the first time, it will prevent that particualr T cell from becoming activated and carrying out its immune function. But it does *not* affect in any way T cells that are not involved in the transplant rejection, leaving them alive and healthy to participate in other immune reactions.

While this exquisite specificity greatly decreases the complications from generalized immune suppression (the AIDS-like consequences), CsA is not without its own toxic side effects. Some are relatively minor, like nausea and the growth of excessive body hair. Of more concern clinically is the *nephrotoxicity* of CsA—its toxicity to kidneys. No one understands completely how this happens, but nephrotoxicity remains to this day a major limitation to the use of CsA.

The discovery of CsA, and its tremendous clinical (and commercial) success, sent drug companies all over the world scurrying to find similar compounds. Remarkably, about six or seven years later another one was found, and even more remarkably, again in a soil fungus—this time in a field right outside the back door of a pharmaceutical company in Japan. This new compound, called FK-506, was first brought to clinical trial in 1989. It turns out to be every bit as effective as CsA, and it is even less toxic to humans. It too works by selectively suppressing the activation of new T cells. And in just the past few years, yet another immunosuppressant, called *rapamycin* (*again,* discovered in a soil microbe), has been cleared for clinical trials. The exciting thing about rapamycin is that it seems to block T-cell activation in a manner completely different from CsA and FK-506, so it may be possible to use these various drugs in combination, at lower strengths for each, reducing the toxic side effects of each.

Thus basic research into immunosuppression has taken us in a few short years from a time when we could transplant only between identical twins, to a point where a wide range of worn-out organs critical to human survival can be transplanted almost at will. Current statistics for some of the more commonly transplanted organs are shown in Table 7.1.

Virtually every U.S. city of more than a few hundred thousand residents now has at least one hospital where transplants can be performed. Not only has organ transplantation received the full backing of the medical establishment, it has—perhaps more importantly—received the approval of insurance companies and Medicare, both of whom are willing to pay for it. Yet this has in turn created a new dilemma. The number of critically ill patients whose lives could be saved by an organ transplant, and medicine's readiness, willingness, and ability to provide one, has now far outstripped the supply of donor organs. This was a possibility not readily appreciated in the heady early days of transplantation, following the introduction of chemical immunosuppression. But it is now the single most important remaining barrier to expansion of organ transplantation worldwide.

Table 7.1. Current Statistics for Some Common Transplant Procedures[a]

Organ	Total number of transplants worldwide	Longest living recipient (U.S.) in years
Kidney	295,000	30
Bone marrow	45,000	24
Liver	27,000	22
Heart	26,000	22

[a] As of 1992.

The Ethics of Organ Procurement: A Modern Moral Dilemma

The magnitude of this problem can be appreciated by taking a closer look at kidney transplantation. In the United States alone, there are over 150,000 people with end-stage renal disease; about 30,000 of these are currently on the waiting list for a donor kidney. Another 12,000 or so are added to the waiting list each year. These patients have no kidney function and cannot clear the poisons produced by their own bodies. They suffer from the same maladies Richard Herrick suffered from, and they will die if untreated. While they wait for a suitable kidney to become available, they are kept alive by kidney dialysis. Although greatly refined since the days when John Merrill introduced it at the Brigham Hospital, dialysis still involves being connected to a machine for at least three half-days each week. The physical and psychological demands placed on such patients are enormous. Their treatments are frequently accompanied by nausea and cramps; patients often develop a negative attitude toward their bodies; there may be transient or even permanent loss of sexual function. Clinical depression and even suicides are common. The vast majority of these individuals do not have a suitable first-degree relative as a potential donor, and thus must wait for a reasonably well HLA-matched cadaveric organ.

Just under ten thousand kidney transplants are performed each year in the United States, from both living related and cadaveric unrelated donors. This number, which has remained relatively constant for the past several years, is limited entirely by the availability of donor organs. As a result, more than twenty-five hundred patients, meeting all the criteria for suitability as a kidney transplant recipient, die from their disease each year while on the waiting list of approved recipients. The average waiting time on this list is now close to one year. Given the tremendous backlog of people kept alive by dialysis, and the stalled rate of kidney transplantation, both the waiting list and the number of candidates on the list who die awaiting a transplant are expected to grow rapidly in the coming years.

For organs other than kidney, the outlook is even more bleak, because the possibility of a living donor does not exist. Moreover, there is no equivalent of kidney dialysis for, say, patients with end-stage heart disease. As a result, although the number of people who need heart transplants is about the same as the number who need a kidney, the waiting list is much shorter for hearts, because people die much sooner after getting on the list. Yet the success rate for heart transplants, when they can be done, is about as good as that for cadaveric kidney transplants. The bottom line is that now, at a time when organ transplantation is more successful than it has ever been before, the number of patients dying for lack of a donor organ is larger than it has ever been, and growing each year.

The solution to this dilemma is clear: Increase the supply of donor organs. Few would disagree with that. The question is how to go about increasing the supply. And that question is at the heart of one of the most important debates in medical ethics in the latter half of the twentieth century.

That debate is still in full force today. Any professional meeting of transplant specialists involving more than a few hundred participants invariably has one or more full sessions dedicated to the ethics of transplantation. Initially, the discussions tended to focus on whether the procedure itself was appropriate. As can be imagined, in the early years of transplantation success rates varied wildly. Both the surgeons and the various postoperative management specialists were engaged in essentially a "learn-as-you-go" enterprise. Many patients suffered considerably, some with little or no benefit. To a person, of course, these were all patients desperately ill—in fact, terminally ill—because of a diseased organ, and so in that sense they had little to lose. But this very element of desperation on the part of the potential recipient itself became a prominent topic in these debates. Can a patient in this situation, or his or her immediate family, rationally analyze the pros and cons of such a new and complex procedure? Medical specialists dedicated to making transplantation succeed in those early days clearly had an agenda of their own, and to pursue their aims they needed patients who required transplants. Were these specialists able to give dispassionate and disinterested advice to potential recipients?

Few would argue any longer that organ transplantation per se crosses any ethical boundaries. Its benefits, balanced against an almost certain fatal outcome in its absence, are simply too compelling. The discussion now centers around what society can or should do to increase the supply of donor organs. For kidney and bone marrow, the donor may be either living or very recently deceased. For all other human organs, recently deceased donors (cadavers) are the only source. Concerning cadaveric donation, some 70 percent of Americans, when queried, indicate that they are in favor of voluntary donation of organs for transplantation. But fewer than 20 percent actually make arrangements before death to do so. How can this gap be closed?

Living Donors. In the case of kidneys, living donors in most Western countries are almost always first-degree blood relatives of the recipient, and these account for roughly 20 percent of all transplants. By more or less common consent, first-degree relatives are considered to have a sufficient personal interest in the survival of the recipient that the slight (but real) risk involved in donating a kidney is acceptable. On the other hand, most transplant centers in the United States will not even consider an unrelated living kidney donor (with the exception of spouses) under all but the most special circumstances. In particular, the notion that a living, healthy person could be *paid* to donate a kidney needed for transplantation has generally been anathema in the United States and most Western European nations. Various regional and international transplantation societies founded in the West, and led largely by American and European scientists and doctors, have repeatedly passed resolutions to the effect that the human body and its component parts cannot be the objects of commercial transactions. Participation in such transactions may be grounds for expulsion from these societies, which also urge governments to make such commerce illegal.

In many developing countries, however, the question of living unrelated donors is approached differently. In India, for example, there are very few kidney dialysis machines, which are expensive to purchase and operate, and there is no organized program for the recovery and distribution of kidneys from cadavers. Yet there are quite a few surgical centers with the requisite trained personnel to perform kidney transplants. Thus the only hope for most patients with end-stage renal disease in India is a transplant. At present, however, the government health system does not pay for organ transplants, and most people do not have private insurance that covers catastrophic illness. Hence, transplants are available only to individuals wealthy enough to have appropriate private insurance or to pay for the procedure themselves out of pocket. In the absence of an organized system for retrieving cadaveric organs, those who have no appropriate first-degree relative, and who can afford it, usually resort to the practice of paying an unrelated living donor to part with a kidney. Inasmuch as the price paid may represent several years' wages to many Indians, there is no shortage of donors. Several thousand such transplants are carried out annually.

Despite expressions of deep concern and even outrage from the industrialized nations of the West, many countries in the same circumstances as India have developed and refined the concept of "rewarded gifting" to get around the injunction against paying living unrelated donors for organs. Most international medical societies that condemn the buying and selling of organs have recognized that living donors are likely to undergo a great deal of personal loss—aside from the organ or tissue donated—in connection with the donation procedure. There may be up to two weeks of costly hospitalization, for example, as well as follow-up treatment and loss of income. Transplant societies generally agree that donors can be compensated for such losses as long as the cost of the organ or tissue itself is not part of any resulting financial transaction. A number of countries have found that by making the terms of such "collateral compensation" sufficiently generous, a monetary value does not have to be put on the organ itself, and the exchange technically does not violate any existing guidelines.

But in fact some developing countries such as India, Iraq, and Egypt have actually gone on the offensive and are working within regional and international transplantation societies to have some form of rewarded gifting officially sanctioned. Some of their points are rather convincing. Listen to Dr. K. C. Reddy, a transplant specialist from Madras, India:

> Those against paid organ transplantation condemn the practice as being "victimization of the poor, a form of corporeal prostitution, resonant with the undertones of slavery." It must be remembered that poverty is one of the grim realities of life in India. Nothing dehumanizes an individual more than poverty, and the inability to provide for one's family. . . . When a mutual transfer, dictated by absolute need, is done with full informed consent of all parties concerned, no serious ethical or moral objection can be made to the act of organ donation for compensation.

Although a statement such as this once provoked a great deal of anger among many Western specialists, in fact the richest one-sixth of the world is coming to realize that the other five-sixths may have a very different point of view about organ transplantation, based not only on economic differences but also on different philosophical and ethical systems. In countries as disparate as Iran and Japan, for example, there is a similar cultural attitude of reverence toward the body of the deceased that makes retrieval of organs from cadavers difficult in the extreme. In Japan, a wealthy nation, very few transplanted kidneys come from cadavers. Patients with end-stage renal disease who do not have a suitable living donor are kept on dialysis essentially until they die. As a result, many transplant professionals are beginning to accept the possibility that with suitable government control to prevent abuse, it may be appropriate for some countries to approach organ donation by living donors differently from what is done in the West.

Unquestionably, if money is involved, the flow of organs will always be from poor individuals to wealthy ones. We in the West find this repugnant. Why? The necessity of asking ourselves such a question was highlighted dramatically by a case several years ago in which a poor Turkish workingman was flown to England and paid $3,300 to donate one of his kidneys. He used the money to pay for an operation needed by his two-year-old daughter, without which she would likely have died. What would the overwhelming majority of us in the West, who would ban organ sales outright, tell this father? That it's okay to donate a kidney to his daughter to save her life, but he cannot sell the same kidney to someone else for the same purpose? By doing what he did he saved *two* lives—his daughter's and that of the recipient of his kidney, who happened to be wealthier than he was. By forbidding his action, and allowing two people to die, would our moral indignation be assuaged?

J. Radcliffe Richards, in a brilliant essay on this topic that touched on the dilemma posed by the case of the Turkish father, had this to say about moral indignation:

> It seems likely that if we forbid [the selling of organs] altogether we shall, for whatever reason, ease our own feelings of disgust. Prohibition may make things worse for the Turkish family and other desperate people around the world, as well as for the relatively rich who will die for lack of kidneys, but at least these people will

despair and die quietly, in ways less offensive to the affluent and healthy, and the poor will not force their misery on our attention by engaging in the strikingly repulsive business of selling parts of themselves to repair the deficiencies of the rich.

. . . If we are forced to recognize that something we find as disgusting as organ selling provides the best option for the destitute and the only hope for the dying, it may help us to keep in mind the need to pursue more radical remedies: on the one hand to increase the effort to find dead donors, and on the other to take the despair of the poor more seriously.

While there may be some willingness to accommodate countries that sponsor and properly supervise paid organ donation from living donors within their own borders, there is still universal repugnance concerning the rapidly increasing and largely unregulated international trade in such organs, which is handled by private brokers on a strictly for-profit basis. According to a recent report of the International Commission of Health Professionals, more than one thousand kidneys from living donors were sold from India in 1988 to various wealthy individuals around the world, mostly in the oil-rich countries of the Middle East, but also in the United States and Europe. Such transactions are not overseen by any official health agency, and they raise a great many concerns. Given that the donor was almost certainly poor and undernourished, what was the state of his (it is almost never a her) health at the time of donation? Could the donor really have stood the rigors of surgery in such a condition? How well are such individuals followed after donation to ensure an event-free recovery? Could consent in such a situation be truly informed? Was the donor screened thoroughly for infectious diseases? The latter has proved to be a major problem in situations where the purveyors of organs recognize no obligation to be involved beyond the procurement itself. In a recent study of wealthy individuals in Oman and the United Arab Emirates who purchased kidneys from India in the mid-1980s, a high proportion were found to have contracted a variety of diseases, including AIDS, through their transplants.

Cadaver Donors. After all the emotion-charged arguments and dramatic anecdotes surrounding the buying and selling of body parts from living donors, discussions of ways to increase the yield of organs from cadaveric donors could be expected to be pretty tame. But in fact the debate on this subject, particularly in the United States and Europe, where living related donors are not even part of the discussion, is every bit as intense as that over live donors. Before we get into the various arguments about ways to increase recovery of cadaveric organs for transplantation, let us look at the sources of such organs.

There are two categories of cadaveric donors in the United States. Many people arrange while they are still alive to make their organs available for transplantation as needed when they die. They do this simply by signing and carrying a Uniform Anatomical Gift Act card, usually pasted to the back of a driver's license.* In such cases, no further consent is needed. The family of the deceased may object strongly to the donation, and physicians and medical centers may respect the family's wishes, but they are not obliged to.

The second category of cadaveric donor includes those who did not make their position on organ donation known while they were alive. In such cases family members of the deceased *must* give their consent before organs can be removed for transplantation. That is the law of the land in the United States, and it is called *required consent*. In some European countries, on the other hand, a deceased person is presumed to have agreed to make his or her organs available for purposes of transplantation after death unless he or she specifically indicated opposition to donation prior to death. This is called *presumed consent*, or sometimes "opting out." However, in practice if relatives of the deceased express opposition to removal of organs from their loved one, this wish is almost always honored.

The controversy over the use of cadaveric donor organs is thus

---

* A valid copy of a Universal Anatomical Gift Act card is included on the last page of this book. If you have always meant to fill one out and attach it to your own driver's license, this may be a good time to do it!

taking place largely in the United States, and in those European countries where required consent governs organ recovery from cadavers. The debate is not about buying and selling organs per se, but rather about how to increase the recovery of transplantable organs after death. Within the context of required consent, this means increasing the number of people who express a willingness while they are alive to make their organs available for transplantation after death. The traditional approach has been to encourage voluntary donations through public education programs. This worked well in the early years of transplantation, but the percentage of the public who carry Uniform Anatomical Gift Act cards has stabilized at about 15 to 20 percent for at least the past decade. How do we get this number closer to the 70 percent or so who say they generally favor organ donation after death but never seem to do anything about it?

One school of thought (for ease of reference we will call them the altruists) maintains that better and more effective public education is the only permissible approach to increasing recovery of cadaveric organs. According to Dr. Renee Fox, a bioethicist at the University of Pennsylvania:

> In organ transplantation, the living parts of a person, offered in life or death to known or unknown others, are implanted in the bodies of individuals in the end stages of grave illnesses. However routinized this human transferral may have become in certain medical and surgical respects since it was first performed almost 40 years ago, it remains an extraordinary act. It is extraordinary because of the literal as well as figurative way in which donors give of themselves, and because it involves surgically mutilating their bodies in order to benefit others. What is given, received, and used in organ transplantation, what it exemplifies and what it transgresses, are all of more than fleshly significance.

There is, from this point of view, something very special about the transfer of an organ from one human being to another, even if the donor is deceased. Above all there is an element of *altruism* in this act, which ennobles both the donor and the act itself, and the altruists insist that this aspect of donation must be preserved. In fact, studies have shown that in the case of living donors, the donor almost invariably does feel very positive about the act of donation, and often experiences an increased sense of self-worth that is long-lasting, even if the transplant fails. Family members who give consent for transplantation of the organs of a recently deceased loved one have similar experiences, often feeling that they have somehow extended the positive impact of the deceased on the world he or she lived in.

While not arguing that transplantation should *not* be performed, the altruists often do contend that the eagerness to transplant may create pressures to supply organs that could lead society onto treacherous ethical ground. There is a sense in these arguments that perhaps we need to rethink the desire simply to use, without question, whatever technologies science can create to prolong human life. At bottom, these proponents would favor an attempt to educate and encourage the public to participate in voluntary, altruistic organ donation, but no more.

This argument seems to strike a chord in most Americans and Western Europeans, probably the same chord that makes us want instinctively to prohibit persons from selling their own organs. But as with the latter question, there is also another viewpoint emerging and demanding to be heard concerning cadaveric organ procurement. It is based on the inherent value of the life of the potential transplant *recipient*. Confronted with the death of any human being who could be saved by an organ transplant, proponents of this viewpoint (whom we will call the pragmatists) find that "poetic statements about the dignity of human life being degraded by commercialism [are] revealed as the empty moral pieties of armchair philosophers incapable of a reasonable balancing of human needs." Strong words, indeed. But these individuals find the ultimate moral repugnance to be the burial or burning of perfectly healthy organs that, if transplanted, could extend another human being's life for ten, twenty, or even thirty years. They greatly resent the implication on the part of healthy, comfortable "armchair philosophers" that someone in the throes of terminal organ failure should simply let go rather than scratch and claw for a chance at a transplant. Let the philosopher speak when he or his child is lying at death's door, they say; then we will listen.

At present only about twenty-five hundred cadavers per year are "harvested" for transplantable organs in the United States, less than 20 percent of the number of suitable cadavers potentially available. While not at all against intensifying efforts to encourage altruistic donation, the pragmatists urge going a step further. They suggest that the reason the number of people who indicate a willingness to donate organs after death does not increase is quite simple: There is no *incentive* for it to increase. Altruism is apparently a sufficient motivating factor for about one in six of us to donate our organs. Another 50 percent or so seem willing to do it in principle, but never seem to get around to filling out a donor card. Any market analyst would immediately suggest that the appropriate corrective would be to provide incentives—modest at first, gradually increasing until the desired level of donation is achieved. Various incentive plans have been proposed and are currently being discussed. For purposes of future reference, we can refer to proponents of this approach as "marketeers."

Nearly everyone favoring this approach agrees that some responsible, not-for-profit intermediary must act as an "honest broker" in any market system for the procurement of cadaveric organs if the system is to gain everyone's trust. This could be the federal government, or one of the numerous professional transplantation societies or agencies already in existence. The basic idea would be to establish a price for various major or minor organs that could be harvested upon death from a suitable donor. In the most optimistic form of this idea, the so-called futures market approach, individuals would be attracted by these incentives to indicate before death their willingness to participate. They could designate the recipient of the financial consideration involved, or specify that it be applied to inheritance taxes, or to pay for funeral or burial costs.

Beyond a doubt, this argument also strikes a deep chord in a number of Americans, and among many Europeans. We use free-market principles to solve problems in virtually every segment of our culture; why would organ donation be any different? We pay people to donate blood or sperm while they are alive; what is wrong with paying their estates for donation of their organs after death? The majority of people who now arrange ahead of time to donate organs at death are financially comfortable and well educated. It is thus unlikely that such a system, if extended, would take unfair advantage of the poor and illiterate. Through an "honest broker" system for distribution, the possibility that the rich would be the primary recipients of the increased influx of organs would be avoided.

Do the altruists buy these arguments? Nor for a minute! Again, Fox offers these observations:

> I am not convinced that permitting a market model will be effective in significantly increasing the number of transplantable organs that are donated. . . . It is neither accidental nor gratuitous that from its inception, human organ transplantation has been based on the belief that "the human body and the extraordinary generosity in the gift of its parts are altogether too precious to be commodified." Because it is institutionalized around the conception of a "gift of life" to serve another . . . it has attained high moral status and transcendent meaning. Its very legitimacy and what it stands for derive from its association with the values of altruism, solidarity, and community. . . . I hope that you will not allow your evangelical faith in the goodness of organ transplantation, and your enchantment with the market, to lure you from your "gift of life" commitment.

There are also serious concerns that any gains made in a market system for organ procurement would be offset by a decrease in altruistic donations. Some people may not want to participate in a scheme so distasteful as the buying and selling of organs. Others may ask why they should give away something that can be sold, and then never get around to selling it, which is bound to be much more complicated than filling out a donor card on the back of their driver's license.

And are the marketeers convinced? Larry Cohen, an economist and foremost proponent of the futures market concept for increasing cadaveric organ donations, has this to say:

> There are those who will view this [idea] as the ravings of a ghoulish market fanatic. They blanch at the thought of a market in so precious and sacred a thing as the human body, even a cadaver. I urge you not to be so delicate and prissy. People are dying while the organs that could restore them to life are being fed to worms. The current prohibition against any and all markets in organs is not rooted in any widespread, deeply felt antipathy to commerce in human organs. It retains its vitality only because those who suffer from it are relatively few in number. . . . Were more to suffer and die from want of the organs that a market could provide, the high-minded pieties that support the prohibition would be revealed for the vacuous moral posturings that they are.

Who would have thought that reaching inside the body and tinkering with the immune system could have created such a storm of human emotion? It is impossible at present to see how these two points of view, so fundamentally different, can ever be reconciled. They represent a basic dichotomy in the human personality that is seen in many, many segments of our culture. The contradiction, as someone pointed out, is not just between altruists and market theorists; it also lies partly in the conflict between our scientific and our cultural conception of our bodies. There is a vague feeling that we cross some invisible yet real line when we mutilate human bodies even for the noblest of reasons, let alone profit. And yet, as the marketeers say, in the midst of the antiphony and cacophony, people whose lives could be saved by an organ transplant are dying, and dying in ever-increasing numbers each year.

It is not easy being human.

Alternatives to Human Organ Transplantation

While the debate over how to increase the supply of human organs for transplantation continues to roll on, physicians and scientists are busy exploring other means of replacing worn-out body parts. Particularly in cases where the use of living related donors is not possible (almost all transplants except bone marrow and kidney), human organs may never be able to satisfy the need even if everyone signed a universal donor card. There may never be enough healthy, transplantable hearts, for example: Too many are defective when a potential donor dies. Lungs are tricky to transplant unless the donor and recipient have chest cavities that are reasonably close in size. Pancreases are notoriously difficult to keep from decomposing in the time it takes to remove them from donors and implant them in recipients. In the sections that follow we will examine some of the alternatives being explored to deal with the inadequate supply of human organs for potentially life-saving organ transplants.

Xenotransplantation. Almost as soon as immunosuppressive drugs made the transplantation of organs between unrelated humans possible, some of the early leaders in this new field began to explore an approach called *xenotransplantation*, the exchange of organs between different species. Perhaps anticipating an eventual shortage of human organs for transplantation, several transplant teams in the early 1960s explored the use of chimpanzee and baboon hearts and kidneys for transplant into humans. They were playing on the hunch that, because these species are so close evolutionarily to humans, there might be a chance for successful transplantation with proper immunosuppression. Moreover, the planned use of a specific animal as a donor would allow thorough advance tissue typing and harvesting of the organ at exactly the right moment for the recipient. These were reasonable assumptions, but early trials were extremely discouraging. Immunological rejection seemed more vigorous than with even the most poorly matched human organs, although one patient transplanted with a baboon kidney did manage to survive ten months. After a few early trials this approach was largely abandoned; fewer than a dozen xenotransplants were performed in the United States through the early 1990s. However, the advent of more potent and specific immunosuppressants such as CsA and FK-506, and the rapidly escalating crisis in the supply of transplantable human organ, has since led to a reexamination of the possibilities of xenotransplantation.

One case that riveted the attention of scientists, doctors, and the public on xenotransplantation was that of "Baby Fae," a female infant born three weeks prematurely with a condition known as *hypoplastic left heart syndrome*. This is a uniformly fatal congenital abnormality in which the left side of the heart is almost completely missing. It affects about one newborn in 12,000, and most of these infants die within a few weeks of birth. Surgery to correct the underlying heart defect is not well developed. Prior to Baby Fae there had been only one organ transplant involving such an infant, using a human infant heart that became available shortly after the unexpected death of a healthy infant. The transplanted infant lived three weeks.

The decision that resulted in Baby Fae becoming the first human infant ever to be transplanted with an animal heart was not taken lightly. Her condition was diagnosed within forty-eight hours of birth by an alert and competent team of neonatal specialists at the hospital where she was born. Both the situation and its inevitable outcome were described clearly and compassionately to the child's parents, who elected to take her home to wait for her to die. On Baby Fae's sixth day of life the hospital contacted the parents—would they consider the possibility of a somewhat radical surgical approach to dealing with their daughter's problem? A medical team at this same hospital had been planning for some time to try xenotransplantation in exactly such a case, and institutional approval had been granted just one week earlier for these trials to begin. After a thorough discussion of all the pros and cons of this approach, the parents approved. The surgeon in charge of the procedure flew back immediately from vacation. Baby Fae was readmitted to the hospital and placed on mechanical life-support while the necessary surgical and follow-up teams were assembled.

The donor was a seven-and-one-half-month-old female baboon. Baboon hearts are remarkably similar to human hearts, and a baboon of the age selected has a heart similar in size to a newborn human being. The transplant took place in a five-hour operation on October 26, 1984, when Baby Fae was twelve days old. This is brutal surgery, but, amazingly, infants are able to absorb far more of this kind of punishment, pound for pound, than adults. Baby Fae came through just fine and was able to feed normally within a few days. As soon as it was clear that she would survive the surgery, and not immediately reject her heart, the hospital notified their regional organ procurement agency to begin searching for a human donor organ. As part of their initial experimental protocol, they viewed implantation of the baboon heart largely as a holding action, what has come to be called a "bridge to transplantation," while waiting for an appropriate human organ.

But Baby Fae would not live to see that fortunate event take place. By the end of the second posttransplant week, her doctors began to detect signs that her heart was weakening. The following week her kidneys began to fail, and finally her heart stopped. After a futile attempt to massage her heart back into life, the youngster died during the evening of November 15, 1984, twenty days after her transplant. At autopsy, Baby Fae's heart showed signs of heavy immune attack, with complications spreading to her lungs and kidneys.

What went wrong? It was clear that immunological rejection had severely damaged the infant's heart, but she had been maintained on levels of CsA that should have prevented T cells from being activated. Like the other higher primates (apes, chimpanzees, orangutans), baboons have histocompatibility antigens virtually indistinguishable from human HLA antigens. The principal mode of immune attack thus should have been by T cells, principally CTLs, with which CsA specifically interferes. In fact, microscopic examination of Baby Fae's heart tissues showed little evidence of T-cell attack. The culprit turned out to be antibodies. Through a bizarre oversight, her surgeons had failed to type both donor and recipient for ordinary blood antigens. Again like humans, the higher primates have standard ABO blood types. Baby Fae was type O, making her a universal donor but unable to accept other blood types. The baboon as it turned out, was type AB—a universal recipient whose blood would be rejected by just about anyone else. There could not have been a worse combination. Enough blood-cell antigens were carried over with the baboon heart to trigger an antibody-mediated rejection reaction by Baby Fae. Cyclosporin A was of no help in this case, because the antibodies needed to reject an AB blood type (baboon or human) were already there. Baby Fae never had a chance; experts wondered afterward how she managed to live as long as she did.

The twenty days of life she experienced with the baboon heart were not much more than the time she might have been expected to live without a transplant. On the other hand, they were days of much better quality than would otherwise be expected. Instead of fighting for every heartbeat and every breath, she was, for a couple of weeks at least, pretty much like any other baby—crying, feeding, and very alert. And she was the longest living human being with an animal heart transplant. The previous record was three-and-a-half days for an adult male human transplanted with a chimpanzee heart in 1977. Although more transplants of this type had been approved by the hospital that treated Baby Fae, the trauma associated with this failure led to their immediate discontinuance.

Recently, a more fortunate outcome has been obtained in the first xenotransplantation to take place since Baby Fae—that of a baboon liver to an adult human patient. It may well be the harbinger of a revolution in organ transplantation. The transplant was performed by Dr. Thomas Starzl, one of the first surgeons to carry out a xenotransplant in the 1960s. He never truly lost interest in it. Over the past two decades Dr. Starzl has established the world's foremost liver transplant center in Pittsburgh, which heretofore had relied exclusively on human cadaver donor organs. As with patients needing heart transplants, a large percentage of people waiting for livers die before one can be found. The recipient in this case was a thirty-five-year-old male whose liver was in an advanced state of failure because of hepatitis B infection. The donor was a fifteen-year-old male baboon who had the same blood type as the recipient. The use of a baboon liver seemed particularly appropriate because baboon livers are not capable of being infected by the hepatitis B virus. Previous attempts to save patients like this man with transplantation of a human cadaveric liver had failed, because residual virus in the body would always infect the new donor liver and destroy it. Thus a major category of patients who could benefit from liver transplantation had been excluded in the past.

The transplant was performed in June 1992, and the recipient lived for seventy days with excellent liver function. He had been in a coma due to liver failure just before the transplant, but was awake and alert within hours after the operation. Five days later he was up and walking. The baboon liver was able to process and store food (one of its major functions) and to make critical blood products like clotting factors. At the time of his death, the patient had high levels of baboon proteins in his blood, and as far as could be determined they were all functioning as well as their human counterparts. Moreover, although the baboon liver was only about a third the size of the patient's own liver at the time of transplantation, by the time the patient died it had grown to nearly normal size for a human.

One of the most promising aspects of this otherwise tragic outcome is that at the time of death the patient showed little or no sign of immunological rejection of his liver. He had been maintained on FK-506 in combination with other immunosuppressants throughout the postoperative period, and this regimen clearly worked. The cause of death was a brain hemorrhage triggered by a fungal infection. The infection was facilitated by a surgical complication that is now understood, and should be avoidable in the future. The patient may also have been maintained on a higher level of immunosuppression than was necessary, which could have contributed to his problems. Although it was a bit premature to draw definitive conclusions, it did not look as though there was any evidence of hepatitis B virus in the baboon liver.

Finally, this case was also rare in that the patient was HIV-positive, although not yet showing signs of AIDS per se. Liver transplantation is major surgery, lasting many hours and with many instruments involved. The risk to the surgical team was considerable. The Pittsburgh team had transplanted a few HIV-positive patients previously, although with normal cadaveric human organs. The possibility of successful transplantation of baboon tissues into human beings raises an extremely interesting possibility for people who are HIV-infected. Just as the hepatitis B virus does not infect baboon liver cells, so too the AIDS virus does not infect baboon T cells. What would happen if an HIV-infected individual were to receive a transplant of baboon bone marrow? Would the T cells that derived from the bone marrow be resistant to HIV infection? Could the baboon T cells, B cells, and macrophages work together to provide immune protection for a human being? Be assured this is a possibility that will be studied very closely, indeed.

Modern Moral Dilemmas: Part II. Whatever clinical problems xenotransplantation may help resolve, it is not going to provide us with an end run around discussions of medical ethics related to transplantation. The participants in these discussions may be different, but the discussions themselves will certainly be no less acrimonious. During the years when no one was doing xenotransplants, all was quiet. The Baby Fae incident almost brought the ethicists up out of their chairs, and it almost stirred the medical community to man the ramparts. But after Baby Fae it seemed that xenotransplantation would go on the back burner again, and no action would be needed. Besides, even the most committed animal rights groups were probably reluctant to argue against saving the life of a human newborn. Part of the strategy for winning a battle is knowing how to select your targets.

But the years between Baby Fae and the Pittsburgh liver transplant were crammed with experiments carried out in many laboratories and medical centers, published openly in the scientific literature, and clearly aimed at exploring to the fullest the eventual use of animal organs in human transplantation. The relative success of the Pittsburgh experiments will certainly drive this program forward, and we can expect to see more xenotransplants in the near future. We can also expect the debate about the moral propriety of these experiments to go up several decibels.

The key buzzword in the new debate is "speciesism"—an awkward word at best but one the public may as well get used to. If xenotransplantation goes forward, we will all hear a great deal more about it. The concept it defines is the following: Does one species—*Homo sapiens*—have the moral right to systematically exploit other species for its own gain? Can the lives of the members of one species be claimed to have an inherently greater moral worth than the lives of other species? The context in which this concept has developed has been stated eloquently by Peter Singer, a bioethicist from Melbourne, Australia:

> Until now the human species, especially so-called Western civilization, has regarded our planet as a resource to be plundered for its own immediate benefit. The animal liberation movement, together with much of the environment movement, is seeking to change this attitude; to get us to see that we share the planet with other species, and that we have no God-given right to exploit them for our benefit. The change is a fundamental one, one that threatens all the major economic forces in our society. . . . It rests on an argument that is so simple, and so plainly sound, that it can only continue to spread.

What does the other side of the debate have to say about all this? In a phrase, "Stuff and nonsense!" For the past ten thousand years at least, they argue, human beings have systematically reared animals for the sole purpose of slaughtering them and eating them. By that act alone, we have already answered the question of whether a human life has greater moral significance than an animal life. How is the proposed use of animal organs for transplantation to save human lives more repugnant morally than using them for food? True, some cultures and a few individuals have opted out of eating animals, but it is clear that human beings evolved at least in part as meat-eaters. The acquired cultural preferences of the few cannot be allowed to suppress the will of the majority.

But a major ethical complication does in fact arise when we contemplate the use of higher primates as involuntary donors in xenotransplantation. ("Higher primates" includes animals such as gorillas, chimpanzees, orangutans, and baboons. Of these, only chimpanzees and baboons have been seriously considered for xenotransplantation.) Numerous studies have shown quite convincingly that these animals are uncomfortably close to humans in more ways than just blood types and HLA antigens. In fact, it is widely accepted that many of these animals in their native state have intellectual and cultural traits as adults that equal or exceed human beings in certain states—any human in the first three months of life, for example, or infants born with severe brain defects. No one would propose using a healthy newborn as a forced organ donor, because that infant is an individual with unlimited potential for development into a sentient human being. But what about an infant born without a complete brain? Even though human, this is a higher primate with less mental capacity than any animal primate, and with zero hope for future intellectual or cultural development. If we would not consider ending the life of such an infant for purposes of harvesting its organs for transplantation, can we in good conscience kill a baboon or a chimpanzee for *its* organs?

This is not entirely a rhetorical question. About one thousand infants are born each year in the United States with a condition known as *anencephaly*, in which most of the brain is missing. Various portions of the head and skull may or may not be present. In cases where a portion of the brain *stem* is present, allowing the infant to breathe, it may live for a week or two, but such infants always die. They never experience pain, thought, or feeling of any kind. To the extent they have eyes, ears, a nose, or a mouth, these are not connected to anything neurologically.

Anencephalics have no knowledge that they are in the world, or even that they are alive.

In a well-known and intensely studied case that occurred in 1992, a Florida couple was told in advance by their doctor that the child they were expecting was almost certainly anencephalic. They decided to carry the pregnancy through to completion anyway, with the plan of offering their child's organs for transplantation into infants like Baby Fae. When it was clear after birth that their infant was extensively anencephalic, they asked the Florida courts to declare the infant brain-dead, the current legal standard by which organs may be harvested before removing all life-support systems. The court demurred, arguing that brain-stem function is part of the definition of brain life. The parents pressed their case through several additional levels of the legal system, arguing that waiting for brain-stem function to cease would result in deterioration of the child's other organs. But the lower court ruling was upheld. The couple's daughter died a week later of lung failure; her other vital organs were unusable for transplantation.

We are faced with a complicated question, indeed. We have declared ourselves unwilling to take the organs from an anencephalic infant to save other human lives; can we then take organs from higher primates? We do not rear higher primates as a food source. If we were to rear them for xenotransplantation, we would be doing so for the express purpose of killing them and removing their organs for transplantation to save a human life. We may well decide to do that, but the ethicists have made a point. If we are not willing to end the life of a human being with no mental capacity *at all* to get organs for transplantation, what is the moral basis on which we believe we have a right to kill a partially sentient higher primate for the same purpose?

The debate over the sale of human organs, and various means to increase recovery of cadaveric organs, has been a bitter one. But it has remained largely cerebral, confined to trading shots through published scientific papers, enlivened by the occasional personal exchange at professional meetings. If the past ten years are any guide, the debate over the use of animals to provide organs for human transplantation will be much more visceral, even physical. More than one university, including my own, has been the target for disruption, vandalism, and personal threats against scientific and medical staff. Clearly, if we are going to go forward with xenotransplantation, we will need the public's support. A case for doing it needs to be made, not assumed, and the public needs to be informed at each step, both about its successes and its failures. We need to answer questions posed by ethicists fully and openly. It is the failure to be completely open and honest about the limitations of animal experimentation, and the ethics of it, that lends it opponents the most effective ammunition.

Molecular Biology to the Rescue (Again!). Bioethicists are not completely united in their opposition to the use of higher primates for xenotransplantation, although most ethicists (and, frankly, many physicians and scientists) are uncomfortable about it. If we could use organs from animals that are already used for food, most serious opposition would drop away rather quickly. The animal most suitable for organ transplantation into humans, after the primates, is the pig. Fully grown pigs are roughly the size of human beings, and their organs are thus capable of supporting the load imposed by human metabolic functions. Hundreds of thousands of pigs raised solely for food are slaughtered annually in the United States, and the vast majority of their major organs end up in sausage or pet food. A few transplants of pig tissues have been tried in humans, but immune rejection has been immediate and violent—a phenomenon termed *hyperacute rejection*. Hyperacute rejection is caused largely by something called *complement*. Complement is used by antibodies to help them destroy invading microbial cells that can cause disease. Most healthy adult humans have antibodies in their bloodstreams that will also attach to the cells of an incoming animal transplant. These antibodies then bind and activate complement, leading to rapid destruction of the targeted tissue.

Complement can be a dangerous substance. In patients with autoimmune disease, for example, or chronic infections that generate antigen-antibody complexes, complement can cause serious problems such as vasculitis (blood-vessel inflammation) or glomerulonephritis (kidney inflammation). These conditions result from physical destruction of cells by complement. Because all of us have a lot of complement running around in our bodies all the time, and because a lot of it gets activated in the ordinary course of fighting bacterial and other infections, our cells are equipped with a variety of devices to protect them against complement damage. An example is a molecule called DAF (*decay accelerating factor*), which promotes the rapid breakdown or decay of any complement molecules that accidentally settle onto normal healthy cells, thus preventing damage. (Complement damage in situations like vasculitis occurs because protective molecules like DAF simply get swamped out by excessive amounts of complement. But under normal conditions, DAF and other protective molecules work just fine.)

Because complement-mediated hyperacute rejection is a principal barrier to transplantation of pig organs into human beings, molecular biologists decided to "build" a pig that was equipped with human DAF. If the heart of such a pig were transplanted into a human being, it would be protected from human complement in the same way normal, healthy human cells are. The technique for making such an animal is one that molecular biologists have used for the past decade to build *transgenic mice*, which have become a routine tool for laboratory research. A fertilized ovum is removed from a recently mated female and the desired gene is microinjected into it. The altered egg is then implanted into a pregnant female's reproductive tract, and in a reasonable number of cases an offspring is born that expresses the *transgene*. The first pigs bearing a human DAF transgene were born in England in early 1993. A number of studies will need to be carried out before the animals are ready to use clinically, and of course they need to be bred into larger numbers if they are to provide a stock for future xenotransplantation. We probably will not see the first transplants of their organs until at least the late 1990s. But if everything goes as planned, the pigs would then be used for the same food purposes for which we currently use pigs, *and* their organs will be available for transplantation.

Artificial Organs. Another option that has been explored for the past several decades is the production of completely mechanical replacement parts for defunct human organs, in a sort of "bionic man" scenario. Serious efforts to design and make such organs started in the 1960s, when the National Institutes of Health, foreseeing a time when the supply of human organs might not be able to meet the need for transplantation, began supporting university and private industry research into artificial organ systems. Work on an artificial kidney had begun during World War II and was by the 1960s fairly advanced. It was already clear by 1970 that an implantable kidney, or even a relatively portable artificial kidney, was not likely to be developed. Biomedical engineers have never figured out a way to reproduce the efficiency of the human kidney on the scale that nature has designed it. And today, of course, kidney transplantation is so successful, especially with living related donors, that work on miniaturization of dialysis machines has virtually stopped.

But an implantable mechanical heart is another story. There are currently almost three thousand patients on the waiting list for a heart transplant. Although some two thousand cadaveric hearts are transplanted annually in the United States, with an average waiting time of about four months, approximately a quarter of those on the list die before receiving a transplant. University, government, and private research teams have worked for the past several decades to overcome the problems associated with creating a mechanical device that could replace the human heart. It is a daunting task. Doctors, scientists, hydraulics experts, naval control systems engineers, oil field drillers, and materials experts have collaborated on electric, nuclear, and air-driven models of this four-chambered marvel. After untold person-hours of effort, and hundreds of millions of dollars of research investment, we are still probably a decade away from the goal of a totally implantable mechanical heart.

That does not mean progress has not been made. There are very impressive heart assistance machines designed to take the load off a failing heart and allow it time to recuperate. These are most often so-called extracorporeal bedside machines to which a patient is connected by tubes. The pump itself is outside the body. These devices were developed by Dr. Denton Cooley, and were used as early as 1969. Depending on the underlying problem, a period of time on such a cardiac assistance device may be sufficient to allow the heart to regain a significant portion of its normal function. If not, and if other factors permit, the patient is removed from the assistance device and placed on the waiting list for a heart transplant. Often a patient may be hooked up to such a machine for a short period after transplantation, to allow the new heart time to settle in before it assumes the full burden of supporting blood circulation in its new home.

Perhaps the best known artificial heart recipient was Barney Clark, a sixty-one-year-old dentist from Washington State. Dr. Clark had been suffering for several years from chronic end-stage heart disease and the associated emphysema. As 1982 drew to a close, his cardiologist told him candidly that he had very little time left to live. He had already been referred to a Seattle hospital for a human heart transplant. Seattle has for many years been in the forefront of American cities developing organ transplant programs; some of the very best surgeons and immunologists in the field practice there. But Barney Clark was turned down because of his age—he was over fifty. Experience at that time had shown that older patients with advanced heart disease simply did not do well with a transplant. The precious few donor organs available were targeted to recipients more likely to receive long-term benefit from transplantation.

However, Dr. Clark had also been in touch with the artificial heart team at the University of Utah headed by Dr. William DeVries. The medical center at the University of Utah was the home of Dr. Willem Kolff, a pioneering Dutch surgeon who had moved to the United States after World War II. Dr. Kolff developed the first artificial kidney to be used in the United States, and in fact had worked with Dr. John Merrill at the Brigham Hospital in the early 1950s. Dr. Kolff had been pushing development of an artificial heart for many years, but unlike those who had gone before, he and his followers were interested in a totally implantable heart, one that would allow the patient to be fully mobile. Research on an air-driven version of the human heart designed by Dr. Robert Jarvik had been intensively pursued at the Utah center. The plan was for the diseased heart to be removed and the entire pump implanted in the chest cavity. The Jarvik heart was not wholly self-contained; the air pumps driving the heart still remained outside the body, connected to the implanted heart by tubes that had to pass through the body wall. Nevertheless, it was designed to allow a fair degree of mobility to the patient. Most of the problems that could be anticipated on paper had been solved using animals into whom the Jarvik heart had been implanted. But it had never been placed in a human before, and the human body puts different demands on a pump of this type than do the animals on which it had previously been tested. These differences needed to be studied, and they could only be studied meaningfully in a human patient. By late 1982 the Utah team was ready for the first human trial, and Barney Clark was selected as the team's first patient. He received his implant on December 1, 1982.

Barney Clark lived with his Jarvik heart for 116 days—difficult days both for the patient and for the team managing him. There were mechanical problems with the heart itself. Parts of the pump did not function as expected. The left ventricle had to be replaced twice. Dr. Clark had to undergo surgical procedures three times to clear up problems associated with his new heart. Only once was he able to get up and move about, and then only with great difficulty. Complicating the purely mechanical adjustments that had to be made was the fact that he had been extremely ill—probably within days of death—at the time of the implant. Many of his physiological and metabolic systems were already seriously compromised.

Dr. Clark's case revealed a poignant but unavoidable side of every bargain underlying medical experimentation: Whenever a completely new technique is tried on a human being—a technique that could, if something went wrong, bring grievous harm to the patient—it can only be tested on someone so ill that there is no other hope for survival. Depending on how it works in essentially terminal patients, the technique may be gradually approved for use in patients less ill.

Even before Barney Clark died, controversy developed over whether the experiment was justified. The medical research community seemed about evenly split on the issue. Those close to the field knew that at some point the only way to find out if an artificial heart would work or not in humans is to try it. The Jarvik heart in various modified forms was subsequently used on a number of other patients, some living for almost two years, and in fact a great deal was learned. Bioengineers discovered how to coat the surfaces of the pump so that clots would not form and bacteria would not attach to it. They gained a good deal of information about balancing pressure and flow once such a pump was placed in a human, and about the design of valves that do not rupture red blood cells.

However, in 1990 the Food and Drug Administration (FDA) decided to suspend use of the Jarvik heart, and the company that manufactured it has gone out of business. The primary reason for its discontinuance was a problem that did not show up in Barney Clark but that did show up in a disturbing number of patients after him: stroke. The cause of these strokes has never been entirely clear. But it was also clear that the Jarvik heart had little appeal to patients and their families because of the limited mobility it afforded. Even in its most advanced version, it still involved a unit about the size of a TV console that had to remain outside the body and be pulled around from place to place. It seems likely now that the air-driven Jarvik heart pump will eventually be replaced by a fully implantable electric heart powered by batteries. Such a heart has shown great promise in calves, and should be ready for clinical trial by the turn of the century.

There is no question that Barney Clark deserves to be remembered for what he did just as much as the doctors and scientists who made his transplant possible. The hundred-plus-days during which he ventured into the unknown were at times filled with intense stress, pain, and physical discomfort. But part of the reason he was selected was that he was enough of a medical scientist himself to know exactly what his condition was from the start, and exactly what he could expect as a result. He also reserved the right to discontinue the experiment at any time should the experience become too stressful. Throughout his ordeal, he showed tremendous courage in facing inordinate pain and suffering. But he also knew that every breath he took, every beat of his artificial heart, provided researchers with information that would make it easier for the next patient, whatever happened to him. Shortly before he died, he smiled weakly and said, "All in all, it's been a pleasure to be able to help people."

Finally, there is one issue that has not yet been squarely faced by those working to develop artificial organs, but which is clearly understood by almost everyone involved. If we are to view artificial organs, or even xenotransplants for that matter, as temporary implants, as a "bridge to transplantation," then we may contribute to human misery as much as we relieve it. For this approach is still dependent for its success on the ultimate availability of human organs. By keeping people alive on an interim basis, we simply increase the size of the waiting lists for the scant number of human organs available. The suffering undergone by these people, most of whom will likely die before a transplant is found, will be for nought. We must truly ask ourselves the question posed by Dr. Denton Cooley, when he was asked to comment on Barney Clark's operation: Are we prolonging life, or are we simply prolonging death?

EIGHT

Minding the Immune System's Business: The Dialogue Between the Brain and the Immune System

The Mind and Disease

Let us return for a moment to something quoted at the very beginning of this book. Thucydides, remember, was writing down his observations about the plague that struck Athens in the fifth century B.C. He is usually recognized by immunologists as having been the first to notice (or at least to record) that people who somehow managed to survive the plague seemed to be protected from ever getting it again. Thucydides was the first to describe what we would now easily recognize as immunity to an infectious disease. But he made another equally remarkable observation, one that I, at least, have never seen quoted by immunologists: "The most terrible thing of all was the despair into which people fell when they realized that they had caught the plague; for they would immediately adopt an attitude of utter hopelessness, and, by giving in this way, would lose their powers of resistance."

The idea that psychological and psychosocial states can influence human health has clearly been with us for a long time, maybe as long as human consciousness itself. It can be found in the literature and traditions of virtually every culture. As the preceding quote makes abundantly clear, it is by no means unique to Western science or medicine. On the other hand, studies in the past hundred years or so have begun to explore these experiences a little more closely, helping us interpret them in terms of what we have learned during that period about how our bodies—and to some extent our minds—function. These studies were the basis for creation of the field of *psychosomatic medicine*; the first professional journal dedicated to this subject appeared in 1939.

Psychosomatic medicine is built on the premise that many disease states arise, at least in part, from the internal responses of individual human beings to psychological and psychosocial events. The responses evoked are as distinct and unique as the individuals in which they are aroused. Many different aspects of the individual are involved—genetic and biochemical makeup, cultural background, and general psychological profile or "personality." All are contributing factors to the way in which any one person may react to a complex of internal and external stimuli. By their very nature, these reactions are not easily reduced to simple "if . . . then" explanations with strong predictive value. One of the triumphs of twentieth-century medicine has been in defining underlying mechanisms of disease in rational and above all *quantifiable* terms. Psychosomatic medicine has not yielded readily to this type of analysis. It is forced to deal with the particular in terms of the general; it is integrative rather than reductionist. It has always stood slightly to the side of mainstream twentieth-century medicine. Physicians and scientists schooled in the "hard" disciplines of biochemistry and molecular biology have always been uncomfortable with the ambiguities of psychosomatic medicine, though few would dispute its major claims.

Mind-related phenomena that can affect human health are many and varied. It has been recognized for many years, for example, that severe personal loss—usually of a spouse or a child—can lead to depression, a sense of hopelessness, and ultimately disease in some individuals. The risk of disease or death among widows or widowers during the period following the loss of their loved one is significantly greater than in closely matched control groups of individuals not experiencing such a loss. Disease in the bereaved, when it develops, is usually more severe and harder to overcome. The origin of the disease-causing disturbance generally lies deep within the emotions and is not easy to quantify, but it is clear that parts of the body, most often the cardiovascular system, can degenerate under the influence of phenomena originating in the central and peripheral nervous systems. Cancer, cirrhosis of the liver, ulcerative colitis, rheumatoid arthritis—all of these and more have been known to develop or worsen during intense grief and bereavement. Conversely, grief-related morbidity and mortality can be *lessened* in individuals who have strong social support systems: family, friends, church, neighbors. This amelioration of the morbid state, as much as its generation, occurs as a result of information processed through the nervous system. Again, the mind/brain can be seen to exercise an effect—in this case a positive one—on the health of other parts of the body.

Similarly, the response to physically or psychologically induced external stess, which is clearly channeled through the mind/brain by our five primary senses, has been shown in numerous human and animal studies to cause dramatic physiological changes in various organ systems throughout the body. These changes are usually related to the need to escape the stress-inducing agent, or *stressor*. In the continuous presence of the stressor, these physiological reactions to stress may permanently damage the organs involved, leading to clear-cut clinical disease. Physiological damage is often most severe when an individual cannot find a way to control or modify the source of the stress. Laboratory experiments with animals make it clear that most of the stressor-mediated changes are caused by hormones of the neuroendocrine system, particularly those like adrenocorticotropic hormone (ACTH) that are involved in the production of steroids. It is assumed that loss-induced disease states have a similar basis. Many organ pathologies are the same in both cases.

But how does the immune system fit into all this? The majority of the literature in psychosomatic medicine has been descriptive, involving the collection of information—often compelling information—that the mind can indeed influence states of wellness or illness. Psychological states affect a wide range of bodily systems. Initially the assumption was that the mind must affect each of these systems directly and independently, but research over the past several decades has suggested that in many instances the mind may have an active partner in this process—the immune system.

The idea that the mind/brain might be affecting many of the body's various organ systems through the mediation of the immune system first gained serious attention in a series of remarkable studies in the 1960s on rheumatoid arthritis (RA). As we saw earlier, RA is autoimmune in nature. Patients with RA make a type of antibody called *rheumatoid factor* that is not specific for an organ or a tissue, but for other antibodies. Aside from the problems this could cause for antibody function, it also leads to the formation of large amounts of *immune complexes*. As the "aggressor" antibodies (rheumatoid factor) collide with and bind to innocent bystander antibodies in the blood, large complexes consisting of antibody sticking to antibody are formed. Normally such complexes are cleared away by macrophages, but when the amounts of immune complex exceed the ability of macrophages to clear them from the bloodstream where they are formed, these complexes may be deposited on the inside lining of blood vessels and, in the case of RA, in the joints. Then T cells, B cells, and macrophages enter the joints, and as these cells try to clear the antibody-antibody complexes away, the smooth tissue that helps lubricate the interaction of bones within the joints is gradually destroyed. This process is painful and, over time, deforming to the joints—the disease we know as arthritis.

Like many other autoimmune diseases, RA has a marked genetic component; it tends to run in families. It also affects predominantly women. But there had been persistent reports in the RA literature that there might be an emotional or "personality" component as well. Patients with RA were consistently described (by their doctors, their family members—and themselves) as "tense," "moody," "high-strung." They tended to have very strict standards for themselves and others, and they reacted negatively when those standards were upset. The problem was that the data on possible psychological elements in RA were difficult to interpret. Data had been collected by researchers in a wide range of disciplines—internal medicine, psychiatry, psychology—each with their own technical approach and particular point of view. Still, a common, underlying theme persisted.

In the face of these intriguing but largely unsubstantiated elements of "common wisdom," Drs. George Solomon and Rudolph Moos of Stanford University's Department of Psychiatry carried out a detailed and carefully controlled analysis of a group of female RA patients. Among other things, they were intrigued by a recent comparison of genetically identical female twins, only one of whom in each instance had clinically diagnosed RA. Clearly in such cases, both twins had identical genetic constitutions; why then did only one sister develop RA? In this particular study, it was found that the twin who developed the disease had had a recent and serious interpersonal conflict accompanied by considerable psychological stress. The authors of the study suggested that development of RA might actually have been caused by an interplay of both genetic and emotional factors.

For their own study, Solomon and Moos chose to analyze not only women affected by RA, but also the nearest-aged healthy female siblings of the RA patient as controls. Applying a wide range of written tests, oral interviews, and clinical examinations, they produced a convincing set of insights into the relation between emotional states and susceptibility to RA. Their data supported some of the previously held notions about this disease, while refuting others. They did not find, as others had previously suggested, that women with RA were more physically active, concerned about their appearance, or dependent in relationships. They did find, however, that in nearly all cases the sisters with RA tended to be more nervous, or more depressed, or quicker to anger in reaction to a real or imagined slight, than their symptom-free siblings. In almost every case, emotional conflict correlated either with the onset or with a pronounced worsening of the disease. Close questioning of the patients and their family members suggested that these traits were not brought on by the burden of the disease itself, but were personality characteristics of the patients before the disease set in.

In a subsequent study Solomon and Moos took a closer look at the healthy sisters of their RA patients. A number of them showed evidence of rheumatoid factor in their blood, suggesting that they may have inherited the same genetic predisposition to RA as their affected sisters. In some cases these levels were even within the range found in patients with active RA. Why then had these women not developed the disease? Psychological testing showed them to be almost exactly opposite in personality type to their siblings with RA. They were generally happy, outgoing individuals who either managed to avoid potentially stressful situations or who coped well with them once they developed. Solomon and Moos concluded from their studies that in some fashion the mind, as manifested in personality, is able to exert a modulating influence on the immune system that can either favor or discourage the initiation or progression of an autoimmune disease, namely, rheumatoid arthritis. This is now a generally accepted notion about the development of autoimmune diseases such as RA, lupus, and multiple sclerosis, among others; onset of the disease may not always represent a failure of the immune system per se, but may reflect a combination of an immune abnormality exacerbated by emotional stress.

These studies show that the mind can exert a direct influence on the immune system itself, in this case helping determine whether or not an autoimmune disease developed. This may be akin to the influence the mind apparently exerts on other specific organ systems—for example, the increase in cardiovascular problems seen in individuals mourning the loss of someone very close. But the immune system is unique among organ systems of the body in that it is instrumental in maintaining *health*. Is it thus possible that the mind exerts an even greater influence on human health by acting through the immune system? Various observations in both humans and in animals suggest this is almost certainly the case.

The most convincing demonstration that the mind—in its perception of and response to stress—can directly influence the body's immune response to a foreign pathogen comes from a very interesting study measuring responses to the common cold. Sheldon Cohen and his colleagues at the Carnegie-Mellon University in Pittsburgh prospectively analyzed 394 physically healthy volunteers to determine their current psychological stress status before deliberately exposing them to a series of cold viruses. Some of the parameters used to evaluate stress levels included recent loss of a close friend or relative; the degree to which an individual felt that current demands in his or her life exceeded the ability to cope; and (as in the RA study) the extent to which a subject described himself or herself with words such as "nervous," "angry," "depressed," or "dissatisfied." Using a composite of all these parameters, the volunteers were grouped in categories ranging from very low to very high stress. Great care was taken to be sure that these categorizations represented the subjects' stress levels at the time of the test and were not generalizations about the subjects' responses to stress at other times in their lives.

After completion of psychological evaluation, and of course after being fully informed of the risks they were about to be exposed to, volunteers were given nose drops containing a low infectious dose of one of five different common cold viruses. They were then housed in special apartments and monitored daily by a physician. Small samples of nose tissue were collected by swabbing to determine whether the virus had succeeded in establishing itself, and each subject was observed closely for standard cold symptoms. The rate at which subjects became infected with the viruses, and the rate at which they developed clinically verifiable colds, *correlated exactly with their stress levels*. For example, 27 percent of the individuals judged to have little or no stress developed colds; nearly 50 percent of those in the high-stress category developed clinical cold symptoms. The rate at which infection and colds developed had absolutely no correlation with a wide range of other parameters such as age, sex, education level, smoking habits, alcohol use, exercise, or sleep habits. This study left little doubt that negative psychological states, and the stress they engender, can weaken the body's resistance to infectious disease, as well as exacerbate internal problems such as autoimmunity.

One of the most famous animal studies on stress and immunity was carried out by Robert Ader in the 1970s. Ader had been using saccharine-flavored water to deliver a potent immunosuppressant drug—cyclophosphamide (CP)—to rats. In both rats and humans, CP is a highly effective inhibitor of the antibody response. Within hours of its administration, the ability to produce antibodies in response to a foreign antigen almost disappears. In the early days of organ transplantation, it was used to help suppress immune rejection of transplanted organs and tissues. In humans, CP induces serious nausea and vomiting, one of several toxic side effects limiting its usefulness in humans. Rats experience virtually the same effects. So nauseating was the CP to rats, in fact, that Ader found the animals very quickly learned to associate the saccharine-flavored water alone with the subsequent onset of nausea and intestinal distress; when he gave them the sweetened water alone, they developed nausea and vomiting as if they had been given the drug. This "anticipatory nausea" would not surprise anyone familiar with Pavlov's conditioning experiments. Dogs who learn that the sound of a bell will be followed by food will salivate (a response normally induced by eating food) just hearing a bell ring.

The anticipatory nausea in Ader's rats looked like yet another case of classical Pavlovian conditioning. Apparently the brain was able from memory to direct the stomach to participate in the same physiological response (nausea) normally caused only by the drug itself. But Ader made a much more startling discovery, one that would fundamentally alter the way in which we think about the immune system. He found that not only did the rats react to sweetened water by replaying the nausea response to CP, but they also developed a state of profoundly suppressed immunity. They exhibited, in fact, the same defect in the ability to produce antibody that is caused by exposure to CP itself.

This was one of the very first demonstrations that the immune system could be manipulated by mental processes alone. The ability to make antibody is one of the most fundamental defenses against invasion by a wide range of potential pathogens. Immune deficiency diseases involving antibody production make clear just how fundamental it is—remember children with Bruton's disease? Ader showed the ability of rats to produce antibody could be deeply affected by a purely mental state—in this case, the rat equivalent of anxiety. He subsequently showed that T-cell responses can be manipulated in the same way. The overriding implication of Ader's studies was that resistance or susceptibility to a whole range of diseases caused by external pathogens, or even internal, opportunistic pathogens, could be affected—*altered*—by the mind.

These revolutionary findings were initially received with considerable skepticism, but they were soon reproduced by other scientists. And the skeptics were unable to comfort themselves for long by supposing that Ader's experiment represented some strange phenomenon restricted to rats; a subsequent study showed an almost identical response in humans. Many drugs (including CP itself) used in chemotherapy for tumors induce nausea and are also profoundly immunosuppressive. In a group of women receiving chemotherapy with such drugs for ovarian cancer it was observed that anticipatory nausea would often begin to develop just prior to a chemotherapy session. It is not all that hard to imagine a wave of nausea overcoming someone about to undergo something one knows from experience will make one sick. But here again, it was discovered that not only were these patients psychologically anticipating the drugs' nausea-inducing effects by developing nausea but they were also entering a state of impaired immune function on their own, well in advance of receiving the immunosuppressive drug. These and other studies of a similar nature showed that, beyond doubt, the human mind has the ability, entirely on its own, to alter drastically the function of the immune system.

Findings like these finally convinced immunologists that the immune system could well be involved in human disease in a much more sophisticated way than they had previously imagined, but they were at a loss about how to approach it from an experimental point of view. The dilemma was admirably summed up by Dr. Robert Good in his foreword to Robert Ader's classic 1981 book, *Psychoneuroimmunology:* "Immunologists are often asked whether the state of mind can influence the body's defenses. Can positive attitude, a constructive frame of mind, grief, depression or anxiety alter ability to resist infections, allergies, autoimmunities or even cancer? Such questions leave me with a feeling of inadequacy because I know deep down that such influences exist, but I am unable to tell how they work, nor can I in any scientific way prescribe how to harness these influences, predict or control them."

Good's statement captures perfectly the quandary that serious scientists of every discipline found themselves in when trying to explore the tantalizing but elusive connection between the mind and the immune system. In his foreword, Good went on to describe an experiment he himself had carried out in 1961 that convinced him such a connection must exist, and which had an interesting twist. Using several individuals who were readily susceptible to hypnotic suggestion, he carried out a series of Prausnitz-Küstner–type allergy transfer reactions. Serum from a highly allergic individual was injected into the skin of both forearms of each subject. The subjects were not themselves reactive to the allergen involved. A short while later, while in a deep hypnotic trance, the subjects were injected in the same spots with a challenging dose of allergen. As we saw earlier in a discussion of allergies, this would ordinarily result in the rapid development of an itching, burning rash and a large welt. Still under hypnosis, the subjects were told that the skin of one forearm was *not* to react to the allergenic challenge, while the other forearm would undergo the normal allergic response. To Good's amazement this is exactly what happened! One forearm developed a severe allergic response, while the forearm that the subject had been told would not respond showed a greatly reduced reaction. Whatever one may think of hypnosis, the fact remains that these subjects were able to control, *through psychological means alone*, the intensity of an immune reaction in their bodies. And again the same questions arise. How? Through what mechanism? How can the mind influence immune reactions at specific sites in the body far removed from the brain?

Explorations of these questions in scientifically rigorous ways that would satisfy both immunologists and psychologists have given rise to an entirely new field of study called *psychoneuroimmunology*. Robert Ader's book of the same title marked a turning point in the attempt to unravel the nature of the dialogue between the mind and the immune system, between brain cells and lymphocytes. Despite its awkward name (no one has yet come up with a better one), this new field has begun to attract immunologists and biochemists, cell biologists, and molecular geneticists, in addition to the experimental psychologists and social scientists who had previously been the primary contributors to the literature. It has required specialists to remove their blinders and look around them, and to become, however briefly, generalists. This is hard for scientists, who become insecure when their delicately arranged platters of knowledge are upset. But this type of temporary disorder has led to some of the most exciting breakthroughs in modern human knowledge.

Lines of Communication

As a way of beginning to understand how the mind and the immune system interact, let us take a quick look at the two principal methods the brain uses to communicate with and control the body generally. The first is by direct *innervation*. That is, nerve fibers from the brain or from the spinal cord can make a direct, physical connection with a particular organ or tissue. By delivering the chemical equivalent of a mild electrical current to a group of muscle cells, for example, the nervous system can cause the corresponding muscle to contract and do work. The chemical message mediating this electrical impulse is one variant of a class of molecules called *neurotransmitters*. But nerve cells are capable of delivering a rather large spectrum of neurotransmitters, the vast majority of which have nothing to do with chem-electrical impulses. Nerve cells can release from their tips a variety of small molecules, often peptides, that bind to receptors on a wide range of nearby cells and alter their activity. There molecules can speed a cell up, or slow it down, or cause it to start or stop producing a specific product that the cell uses internally or exports to other cells. Among this potpourri of neurotransmitters are such molecules as nerve growth factor (NGF), whose normal function is to stimulate nerve growth, and vasoactive intestinal peptide (VIP), which causes blood vessels to dilate or contract. All of the immune tissues such as thymus, spleen, and lymph nodes are extensively infiltrated with nerve fibers from the brain and spinal cord. Clearly, the immune system is in *potential* connection with the immune system through direct innervation and local release of neurotransmitters.

The second major way the brain exerts control over bodily processes is through chemical messages exchanged within the *neuroendocrine system*—an interactive network involving distinct regions of the brain and certain endocrine glands scattered throughout the body. Direct hard-wiring of the brain to body tissues through nerve fibers is easy to visualize. The neuroendocrine system is a more complex and subtle communication network; the way it works is outlined in Figure 8.1 (p. 234). Two distinct but intimately related structures in the brain are involved in chemical communication with the rest of the body: the *hypothalamus*, which sits just about in the middle of the brain; and the *pituitary gland*, suspended just below the hypothalamus. The pituitary gland produces six key peptide hormones used by the brain to run its neuroendocrine communications network:

Luteinizing hormore (LH) and *follicle-stimulating hormone* (FSH) regulate the female reproductive cycle.

Growth hormone (GH) is important in stimulating and regulating the growth of soft tissues and bones.

Thyroid-stimulating hormone (TSH) regulates function of the thyroid gland.

Prolactin (PL) stimulates milk production during pregnancy and subsequent nursing.

Adrenocorticotropic hormone (ACTH) causes the adrenal glands (located just above each kidney) to produce corticosteroids, which prepare the body for "fight or flight" but which may also be involved in stress reactions.

The hypothalamus, which receives and integrates information from various other parts of the brain, regulates the release of these six neurohormones from the pituitary largely through the production of so-called *releasing hormones*, also shown in Figure 8.1.

The important difference between control of body functions through direct innervation and control through the neuroendocrine system is that in the former case a message is delivered to and released at a very specific site; only cells within a millimeter or so of the message are likely to be affected. In the latter case, a chemical message is simply dumped into the bloodstream, where it circulates throughout the body and can potentially affect any cell anywhere in the body that has a receptor for that message.

It had been known for many years that stimulation or destruction of various regions of the brain could have profound influences on immune reactivity in animals, affecting both antibody and T-cell responses. This certainly implied some form of communication between the brain and the immune system, but it was always viewed as possible that such effects were indirect. For example, the brain could affect tissue A, which influences organ B, which then has some impact on immune reactivity. But in the early 1980s scientists began to unravel the ways in which the brain (and thus the mind) communicates with the immune system. To

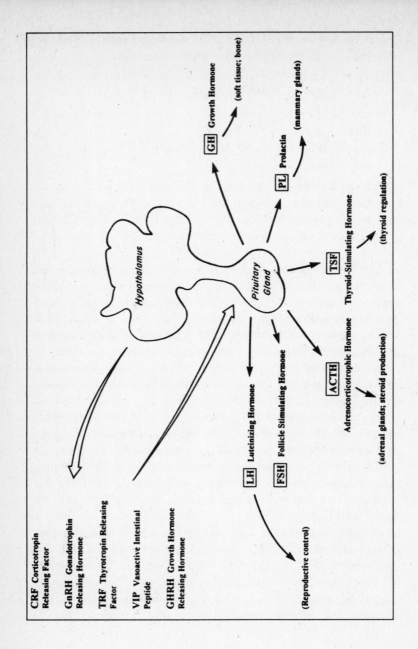

CRF Corticotropin
Releasing Factor

GnRH Gonadotrophin
Releasing Hormone

TRF Thyrotropin Releasing
Factor

VIP Vasoactive Intestinal
Peptide

GHRH Growth Hormone
Releasing Hormone

Hypothalamus

Pituitary Gland

GH Growth Hormone

(soft tissue; bone)

PL Prolactin

(mammary glands)

TSF Thyroid-Stimulating Hormone

(thyroid regulation)

ACTH Adrenocorticotrophic Hormone

(adrenal glands; steroid production)

LH Luteinizing Hormone

FSH Follicle Stimulating Hormone

(Reproductive control)

234

everyone's delight, if not surprise, it was found that cells of the immune system, like many other cells in the body, have surface receptors for both neurotransmitters and neurohormones. They are thus prepared to receive messages from either of the brain's two broadcasting systems.

And it is clear from recent experiments that cells of the immune system are exquisitely sensitive to the brain's messages. Both T cells and B cells can be isolated outside the body and made to respond to antigen under strictly controlled conditions. When such cells are exposed to neurochemical signals in combination with their exposure to an antigenic stimulus, profound alterations in their responses can be observed, and these alterations occur in response to levels of the neurochemicals actually found in human blood. The immune response can be enhanced or suppressed, depending on the particular neurochemical or combination of neurochemicals used, and the timing of their administration with respect to exposure to antigen. The receptors on immune cells for these neurochemicals seem to be identical with receptors found on other cells, and neurotransmitters and neurohormones probably affect T and B cells in the same general way they affect other cells: They speed them up, slow them down, turn genes on and off, and affect synthesis and secretion of cell-specific products.

The really surprising finding to come out of these studies is that not only does the immune system have specific receptors that allow it to receive chemical messages produced by the brain but the brain also has receptors that allow it to monitor the lymphokine messages exchanged among white blood cells. The T helper

FIGURE 8.1. (*facing*) *The symphony of neuroendocrine messages directed by the hypothalamus-pituitary region of the brain.* The neurohormones shown in the box at the left are produced by the hypothalamus, and in turn induce production of the six major neurohormones of the pituitary gland. Essentially all of these neuroendocrine messages are also made in the immune system and can be received and interpreted by immune system cells.

cells produce a wide range of chemical messages called *inter-leukins* to orchestrate activities in other cells of the immune system. Some of these, like interleukin-2 (IL-2) help B cells to make antibody; others (IL-2 plus IL-4) assist killer T cells in their maturation phase; still others (IL-1, IL-6) are involved in general inflammatory responses. The brain has receptors that allow it to tune in on this immune system conversation and to interpret what is going on in the body at any given moment with respect to immune defense.

Immune system messages are not simply read and filed away by the brain. These messages can in fact have very specific and profound effects on brain activity. It was known for many years that immune reactions to foreign invaders could alter brain cell electrical activity, which in turn could affect, for example, neurohormone release from the pituitary gland. An interleukin such as IL-1, made by B cells and macrophages, can also tell the hypothalamus to raise the body's temperature. This causes fever, which helps the immune system fight infection. At the same time IL-1 increases pituitary output of ACTH while reducing production of LH, FSH, and prolactin. This makes it harder to start a pregnancy during strong immunological responses to infection. Also, IL-1 acts on other parts of the brain to enhance sleep and to decrease appetite and digestive function. Other lymphokines are thought to cause the brain (through direct innervation channels) to alter blood flow near the site of the infection, making it easier for immune cells to reach the site. Thus the principle of two-way communication between the brain and the immune system was firmly established.

This communication process is not without its perils, however. Excessive levels of immune system signals coming into the brain can wreak havoc with mental function. Because of its potent stimulatory effects on T cells, IL-2 has been given to cancer patients to bolster their immune defenses against their own tumors. But it was found that high levels of IL-2, among other things, caused serious mental disturbances in these patients. Although these effects reversed when the drug was stopped, its induction of such neuropsychiatric disorders is now recognized as a major limitation of the use of IL-2 and other lymphokines in immunotherapy for tumors and other medical problems. In the case of AIDS, where the T cells run amok and produce excessive amounts of nearly all lymphokines, the resulting "chemical cacophony" puts unbearable stress on the brain and contributes to the AIDS dementia complex that is often a devastating manifestation of AIDS.

The story of the pathways used in the dialogue between the brain and the immune system would take one final turn. A young man with an impressive beard and the name J. Edwin Blalock arrived at the University of Texas Medical Branch at Galveston in 1976 to begin his postdoctoral studies in preparation for a career in research. New PhDs in science are almost never hired directly into the best academic or industrial research positions without an additional two or three years' seasoning as a "post-doc." The postdoctoral years are the period when one demonstrates an ability to be productive and to do independent and creative science. It is a time for becoming known to the larger world of scientists and for building the network of colleagues and future collaborators that are the backbone of every scientific career. It is also the time when one starts to focus in on the specialization within science that may become one's life work, or at least one's work up through tenure—which is about as far as life seems to extend at that stage.

Ed Blalock's focus for the future had turned to lymphokines, and in particular a subset of lymphokines called *interferons*. When cells, including lymphocytes, are activated by an encounter with a virus, they release large amounts of interferons, which alert surrounding cells to the presence of a viral prowler in the neighborhood and help them get ready to protect themselves from viral invasion. Interferons isolated from activated lymphocytes can be purified and used clinically to help fight serious viral infections, so there was a lively interest in their properties, and especially in their isolation.

Blalock was intrigued by reports that highly purified preparations of one of the interferons (*interferon-alpha*, or IFN-α) seemed to act like the neurohormone ACTH when mixed with ACTH-sensitive cells. Other people who had noticed this more or less assumed that IFN-α must be accidentally cross-reacting with cell-surface receptors for ACTH, causing them to react as if stimulated by ACTH itself. Blalock did not like this idea. Nature is rarely so sloppy. Cytokine signaling between cells is (has to be) very precise. IFN-α has one message to bring to a cell; ACTH quite another. But the alternate explanation—that lymphocytes were making and releasing a cytokine produced by the brain—was equally hard to believe. Why would cells of the immune system be essentially "forging" messages that should be sent only from the brain? This was indeed a problem worthy of investigation. This was a question that, depending on the answer, could indeed become a lifetime's work.

The answer came much more quickly and clearly than even the most earnest post-doc could have hoped for. Applying a battery of tests based on antibodies and enzymes, Blalock proved that lymphocytes were indeed making and releasing into the bloodstream *exactly* the same ACTH molecules made and released by the pituitary gland in the brain. He quickly wrote up his findings and published them in the prestigious journal *Proceedings of the National Academy of Sciences*. This opened a completely new era in the study of mind–body communication. The immune system is able to understand chemical messages coming from the brain, and the brain has also learned to understand "immunese." But now we suspect that the immune system has also learned to speak the same language the brain uses. Further research has provided evidence that the immune system is able to make over twenty neuroendocrine molecules thought previously to be the exclusive domain of the brain. Many of these molecules have now been analyzed structurally and shown to be identical to their counterparts produced in the brain. And to complete the cycle, it has recently been found that the brain itself is able to make the immune system messages IL-1, as well as IL-4 and IL-6. Not only does the brain understand "immunese" but it has actually learned to speak it. So in reality, rather than sharing a single, common language, it turns out that the mind and the immune system are completely fluent both in their own language and in the language of the other.

Does each speak the other's language loud enough to be heard, to cause the other to stop or go or turn back? We cannot be completely sure yet, but recent experiments have shown, for example, that the β-endorphin produced in lymphocytes, which is apparently identical to the same painkiller produced in the brain, is able to quell the pain response in peripheral nerve fibers. It has also been shown that lymphocytes deprived of their endogenously produced supply of GH (growth hormone) do not proliferate normally in response to growth signals. It is getting increasingly difficult to imagine that brain chemicals produced by lymphocytes are just an intriguing but meaningless coincidence!

Talking It Out

Beyond any shadow of a doubt, there is bidirectional communication—a genuine dialogue—between the brain and the immune system. It is almost as if the immune system were a chemical extension of the brain floating around in the bloodstream. But why is it that among all the physiological systems governing how our bodies function, only the immune system and the brain speak and understand each other's language? Obviously, the various targets of the neuroendocrine hormones, such as the thyroid gland or various reproductive organs, receive and interpret messages from the brain. But they do not talk back; they do not themselves produce brain-associated neuropeptides. Nor does the brain itself make any of the chemicals produced by these target organs.

And why is it that in the discharge of their respective functions, only the nervous system and the immune system, among all the physiological systems in the body, share the property of memory? Only the brain and the immune system can store information based on previous experience and use that information to alter responses to future events.

It may be that the brain and the immune system need to speak each other's language because they share a common function: constant surveillance of the world in which we live. Knowledge of the world outside our bodies is conveyed to the brain by our five primary senses—sight, sound, touch, taste, and smell. We use these senses to find food and shelter, to recognize our own kind, to guide our bodies through the three-dimensional maze we call our world. But we also use these senses to perceive danger, to detect threats to our physical well-being. Our senses relay this information to the brain, which converts this information into chemical and electrical signals that we use to overcome or to avoid the perceived danger.

As Blalock has pointed out, the immune system functions very much like a sixth sense. But it is a sense with a unique mission. The five primary senses are all directed outward, probing the world around us for information crucial to our survival. The immune system surveys the world within us. It is able to sense the presence of potentially threatening invaders like bacteria or viruses, or of newly emerging tumors, none of which can be detected by the five primary senses. But the immune system must be able instantaneously to translate what it senses into information the brain understands—into a language the brain itself uses to regulate body functions. And so over time the mind and the immune system have learned to talk to each other in terms each understands. The implications of these new findings for biology and medicine are only beginning to be appreciated, but they hold the promise of answers to questions as old as medicine itself.

So where do we go from here? No one would now question that the mind can accelerate, or possibly even initiate, morbid conditions in the body, and it is now abundantly clear that often this is mediated through the immune system. We need a lot more research at the chemical and molecular level to understand exactly how this works, and when and where it is important. But must we focus only on the negative? Is there also a possible upside to this story? Is health simply the passive absence of disease? Many studies have suggested that strong, supportive personal and social relationships correlate with good physical health. Unfortunately, a lot of this information is purely anecdotal at this stage; more research is sorely needed. But could it just be that the mind, in addition to causing disease, can also help us actively ward it off? How would that work? Would the immune system be involved? How would we use the information that the mind can work either positively or negatively, vis-à-vis human health, to our advantage?

Scientists have only been listening to the cross-talk between the brain and the immune system for a few years. We have much to learn about how these two crucial systems work together; about which system dominates in a crisis; about who decides to listen to whose messages first. One thng is clear: These two systems are extensively interconnected. Any alteration of messages in one system is likely to have profound and possibly unforeseeable effects in the other. Thus, we are not yet ready to try to intervene in this communications network; indeed, we may eventually decide that doing so would be totally inappropriate. We will probably have to listen in for many more years before we can even make that decision. But the code has been broken, and we are listening!

Diversity, Tolerance, and Memory:
The Politically Correct
Immune System

The history of immunology presented in the early part of this book was, for the most part, a social and political history of immunology. Until the very end of the nineteenth century, immunology did not exist as a distinct field of study, and thus did not have an intellectual history. Only with the discovery of antibody in the blood of immune animals would the great scientific debates defining the science of immunology in the twentieth century be joined. This epochal discovery was made in the laboratory of Robert Koch and first published in 1890. The initial scientific discussions of the significance of this discovery were very much colored by the intense political and nationalistic rivalries between the Koch and Pasteur camps described earlier. But these debates were no less valid or convincing for having been driven by passions surviving from the Franco-Prussian War. Quite the opposite. Throughout, both sides held to the highest scientific standards in the experiments they designed, executed, and interpreted. Part of the reason for such high standards was doubtless the knowledge that a group of very talented scientists on the other side of the political line would be examining every pronouncement with the utmost suspicion and skepticism, so it had better be absolutely air-tight. But the bottom line was that a lot of very high-quality science was done in a very short time, and the science of immunology was launched on a solid footing.

Once the polarization between the French and German camps died down, and contributions from other scientific centers in Europe and North America started to pour in, immunology finally did begin to generate an intellectual history of its own. The unraveling of the inner workings of the immune system has resulted in enormous dividends for human health, affecting almost every branch of medicine. But the story of the *process* of this unraveling is in itself one of the most exciting intellectual dramas of our time. The important events in this history, as it evolved, were driven more by purely intellectual rivalries than by nationalistic feelings, but the rivalries were not necessarily any less intense. Science of any kind is intensely competitive, and the driving force for this competitiveness is at bottom the human ego. The vast majority of scientists who have moved scientific knowledge forward most effectively have had one thing in common, regardless of their national origins. They wanted to be right, and they wanted to be right *first*.

The theme of the great debates about immunology that began at the end of the nineteenth century, and from which sprang the true intellectual history of the field, had a single goal: To try to understand what the immune system is and how it works. The task set for the immune system is clear: to protect us from invasion and infection by microbial life-forms that threaten our lives. How does it go about doing that? What are the key challenges to accomplishing this task? What are some of the inherent limitations under which the immune system must function?

An understanding of the immune system and how it functions requires above all an understanding of the relationship between the human body and the enormous numbers of pathogens that would like to live in it. The conditions that evolved within our bodies to sustain human life are not very different from those that sustain life and promote reproduction for bacteria, funguses, parasites, and viruses. Microbes have exploited every niche on this planet that can even remotely support their existence. Compared with a boiling sulfur vent at the bottom of the sea, or the freezing temperatures at the edge of the arctic tundra, human bodies are not a bad place for microbes to work and play. A limited number of microorganisms cohabit quite successfully with humans, and in fact help us to carry out a number of tasks we cannot do on our own. Bacteria in the gut, for example, help us to digest certain foods we would otherwise simply pass through with no nutritive gain. In exchange, we instruct our immune systems to leave them alone. Other microbes that are potentially pathogenic are kept under control (although not completely eliminated) by the immune system. This allows them to survive and reproduce within us, although at levels too low to harm us. Most of the time this is acceptable, although it is a situation that can sometimes result in serious problems, as we have seen in the case of opportunistic pathogens. But a great many microbial life-forms that range from mildly to violently pathogenic have the ability to invade and establish themselves in our bodies. They are what our immune systems must destroy, before they destroy us.

Microbes, of course, have their own agendas, which do not necessarily include our own personal well-being. Like all living things, their principal goal is to reproduce and pass on their genes to the next generation. Many microbes have learned to live in harmony with their host, so they can take full advantage of what the host has to offer without killing it and being forced to look for someone else to sponge from. But for many microorganisms, it does not seem to matter much whether the animal or person they live in is alive or dead. It would be nice if all microbes were rational, and would avoid destroying a perfectly good host, but microbes are not equipped with reason, and in the end it isn't really all that diffcult for many of them to get from a dead animal to another living one.

From an evolutionary point of view, one single fact above all others dictated the final shape of our immune systems. All mammals, including humans, evolved in a world full of microbes that adapted to our flesh as we struggled to establish our own niche in the biosphere. During this period of co-evolution, we adapted strategies to limit their advances, while they developed means for evading them. But this was far from an evenly matched contest. It is extremely important to understand that genetic evolution takes place in the time between generations, the time when the genetic deck is shuffled and new hands are dealt. Some of the changes that arise in this fashion are selected by the environment to succeed, and some fail. This is equally true in animals and in bacteria. *What is not equal is the time frame in which these changes take place.* Bateria may deal a new hand (that is, reproduce) every thirty minutes, when all is going well; humans, even at their most prococious, require at least a dozen or so years. Thus the advantage for genetic variation as a means of problem solving is far and away to the bacteria (or yeast, or parasite, or virus). Microbes can try, and discard, a hundred thousand strategies for evading our defenses in the time it takes us to see if our last hand was even playable.

Thus the first problem the immune system must solve is how to deal with microbes that can, through genetic mutation and variation, evolve evasive strategies far, far faster than animals can evolve defensive strategies, through standard genetic variation, to counter them. This problem becomes particularly acute in evolution with the appearance of the vertebrates, the animal group to which humans belong, and in which the time required to reach sexual (reproductive) maturity begins to lengthen. It is the length of this pre-reproductive period that determines that time between generations. It is probably not a coincidence that it is only with the vertebrates that the immune system as we know it first makes its appearance.

The way vertebrates deal with the ability of microbial predators to mutate so rapidly is not subtle, but it is simple and very, very effective. It relies on the time-honored strategy of brute force. We can make so many different kinds of antibodies that no matter in which direction a pathogen mutates, we have an antibody (or a T cell) to block it. We have evolved a system for mutating our immune system genes internally, *independently of our own sexual reproduction*, at rates that far exceed the mutational rates of the microbes that threaten us. Moreover, we can generate this diversity in an ongoing, daily fashion at a rate much faster than any microorganism can reproduce. This process is perhaps the most important evolutionary development in the immune system. It allows vertebrates to generate literally millions of different kinds of different kinds of antibody every day. As we shall see, this is done blindly, without any knowledge of what the antigenic universe might look like, in particular without any knowledge of what a pathogen might look like. This feature of the immune system is referred to as *immunological diversity*.

When we think about it, vertebrates had no other choice. Microbes have the ability to change themselves by standard genetic means so rapidly, compared to vertebrates, that any genetic changes made by one generation of vertebrates in response to pathogens it had recently encountered would most likely be useless in succeeding generations, because of the ability of microbes to alter themselves so quickly. Not only must the immune system be able to deal successfully with all currently existing pathogens, but with pathogens yet unknown, which can develop in much less than the time it takes to generate a human. For example, the virus that causes AIDS (HIV) changes its antigenic properties so rapidly that an individual infected with one form of the virus will end up producing several immunologically distinct variants of that virus *within his or her own lifetime*. Basically, the vertebrate immune system has given up entirely on trying to compete with microbes on their own terms, and has developed a strategy that is completely independent of reproductive mutational rate. We produce such a huge variety of different antibodies that we are able to cover what seems like an almost infinite antigenic universe.

An appreciation of how vertebrates meet the challenge of rapidly evolving microbes did not come easily. Although seemingly of little immediate practical interest, this question nevertheless tested some of the best minds in immunology during the first seventy-five years of this century. Most early immunologists spent their time, in the fashion described so well by Thomas Kuhn, injecting ever more microbes into rabbits or horses or guinea pigs and studying the properties of the antibodies that were invariably produced. The goal of these experiments was to develop antibodies and immunization strategies for use in treating human disease. We are all of course glad that they did this, but there were always a few scientists who were less interested in the antibodies themselves than in how they are produced. Where do the antibodies come from? How do they get into the blood? Are they there all along, and just called out by antigens? What are they doing when there is no antigen around? How is it that the body is able to produce antibodies against virtually any pathogen, antibodies that recognize the immunizing pathogen and no other? These questions are at the very core of immunology. Attempts to address them would shape experimental approaches to understanding immunology for the next hundred years.

One of the first to put forward a possible model for antibody production was the brilliant German physician-scientist Paul Ehrlich. Ehrlich brought a broad background in medicine and chemistry with him to the new field of immunology. He also brought an extraordinary intellect and the gift of making other scientists think, often by proposing something so outrageous they had no choice but to try to think of a better alternative. They rarely could, but at least he got them to try. In addition to his seminal work in microbiology and immunology, Ehrlich also developed the first truly effective treatment for syphilis, a drug called Salvorsan. Despite receiving a Nobel Prize for his immunological work in 1908, the Nazi government in Germany later tried to destroy almost everything Ehrlich had accomplished, simply because he was a Jew. Fortunately, Ehrlich was spared this humiliation; he died in 1915, honored and deeply respected in almost every country in the world—including Germany itself, before the ascendance of the Nazi era.

Ehrlich first became interested in immunology through his interest in the interaction of bacterial toxin molecules with antitoxin antibodies. He approached this as a problem in chemistry, a still relatively novel—and controversial—idea at the time. Ehrlich was a great believer in the idea that biology is in the end just a fancy sort of chemistry. As can be imagined, this was not readily accepted by many late nineteenth-century romantics, although scientists of the period were increasingly drawn to this notion. Ehrlich certainly was one of the first to introduce a quantitative approach to the study of immune reactions in vitro ("in glass," i.e., in a test tube outside the body), and he is generally regarded as the founder of the field of immunochemistry. Chemistry in the late nineteenth century was a well-established, highly productive, and promising scientific discipline. Immunology appeared to be a useful clinical subspecialty but had not established any scientific credentials of its own. Thus, many regarded immunochemistry as the only truly scientific aspect of immunology for the entire first half of the twentieth century. Attempts to study the immune system in the whole animal had revealed very little about underlying mechanisms. Although extraordinarily successful in treating disease, immunologists did not have a clue as to how it all worked in vivo (i.e., in living animals).

Ehrlich also became interested in the problem of where antibodies come from in the body. In 1900 he have an address before the Royal Society in London in which he laid out a theory he had been developing for the past several years. He made the novel suggestion that antibodies might be related to cell surface receptors for nutrients. He imagined that in order for cells to take up nutrients from their surroundings, there must be specific receptor molecules at the cell surface, binding the nutrient to the cell and facilitating its entry into the cell. He imagined different receptors with chemical specificity for different nutrients. A mature organism would have to have a wide range of such receptors to take advantage of the wide range of nutrients in its environment.

Key to Ehrlich's theory was the notion that some of these nutrient receptors would either directly recognize, or cross-react with, various parts of microbes or their toxins. This would result in the toxin being bound tightly to the cell and would explain, incidentally, how toxins penetrated cells and killed them. But in terms of the immune response, Ehrlich proposed that when cells are repeatedly stimulated through their nutrient receptors, they would overproduce copies of these receptors/antibodies and shed them into the bloodstream. This was in effect a *selective theory* of antibody production. The complete spectrum of antibodies (nutrient receptors) and the cells that produce them are already there; antigen coming into the system merely selects some of them for production and release.

As we will see, Ehrlich's model would prove to be amazingly close to the truth, but a decade or two after he proposed it, it received a setback that would take many decades to overcome. What Ehrlich's theory seemed to imply was that all antibodies preexist in the animal without any prior knowledge of the antigenic universe per se. Since he envisioned the primary function of antibodies as nutrient receptors, their ability to act as antibody against pathogens would have to depend on chance cross-reactivity of pathogens with nutrients. The nutrient cross-reactivity notion can be stretched quite a bit, and in any case could not easily be disproved. But it ran into very serious trouble with the work of Karl Landsteiner, an organic chemist who came from Austria (via the Netherlands) to the Rockefeller Institute in New York in 1923. In addition to his contributions to immunochemistry, he produced an enormous body of work with far-reaching implications in other areas. It was Landsteiner, for example, who first defined the human blood cell ABO groups, making blood transfusion a clinical reality. He received the Nobel Prize for this work in 1930.

But as an organic chemist, Landsteiner had a passion (and a talent) for tinkering with the structure of molecules, altering them ever so slightly and observing the changes in their properties. Intrigued by the chemical specificity of the interaction of antibodies with their antigens, he began altering the molecular structure of antigens to see which changes in structure would affect their ability to interact with the antibodies raised against them. What he found was that every time he made a small change in an antigenic molecule, the body would respond to this change by producing a slightly different antibody to it. Thinking back to Ehrlich's model, it dawned on Landsteiner that some of the molecules he was producing were very unlikely candidates for nutrients. Many were exotic organic compounds far removed from the food chain. In fact, some of the molecules he produced in the laboratory did not exist in nature, yet could stimulate antibody production. Why would cells have receptors for nutrients that had never before existed on the face of the earth? In the minds of most scientists, including immunologists, Landsteiner's findings simply could not be reconciled with Ehrlich's nutrient receptor theory, and Ehrlich's theory was ultimately abandoned.

Nevertheless, the problem of where antibodies come from, and how they come to be so specific for the antigens that induced them, continued to intrigue immunologists and chemists alike. After Landsteiner's work, most theories assumed that antibodies must somehow be formed in the presence of, and under the direct physical influence of, antigen itself. Although a number of thinkers contributed to the shaping of these *instructive theories*, perhaps the most polished version was put forward by Linus Pauling, the Nobel Prize-winning (1954) chemist from Cal Tech, in a paper published in 1940. Pauling shared with Ehrlich the ability to provoke people to think, often by proposing something bordering on the outrageous—but always, of course, sufficiently close to accepted dogma/common wisdom that it had to be taken seriously.

The premise of all the various instructive theories, which had great force and influence on the thinking of immunologists until the late 1950s, was based on a false assumption about proteins. It was known by the 1930s that antibodies are proteins. Pauling and the other proponents of instructive theories of antibody formation assumed that protein structure (i.e., protein three-dimensional conformation) was a plastic property of proteins, one that could be altered or *instructed* by contact with another molecule. What they imagined was that antibodies might be produced as essentially blank templates that, by interaction with an incoming antigen, could be molded into a complementary shape that would have the ability to interact in a high-affinity fashion with other molecules of the same antigen. This would be something like having blank keys that, when inserted into a lock, would assume the shape of the lock's interior. Such a model could account very nicely indeed for the specificity of antibodies, since they would assume shapes that fit exactly around the incoming antigen. Most important, it got around the awkward notion proposed by Ehrlich that the entire repertoire of antibodies existed completely pre-formed, with no prior experience of antigen. The instructionalists proposed that the immune system contained no a priori information about antigenic molecules in the environment, but rather learned about them only as a result of their first physical contact with them.

Although offering instant relief from the anxieties presented by Ehrlich's selective theory and Landsteiner's challenge, the instructive theories had problems of their own. One of the earliest difficulties to be pointed out was the fact that there was no means proposed in these theories to account for *tolerance of self*. If antibodies form simply by shaping themselves by contact with antigen, then why wouldn't an antibody form itself by wrapping around self molecules, as well as foreign molecules? Why wouldn't there be antibodies to both self *and* foreign molecules? No one could offer an answer to this problem, but it was assumed, as is often the case when the challenge is largely theoretical, that an answer would somehow make itself apparent somewhere down the line.

A second shortcoming of instructionalist theories was the inability to provide a satisfactory explanation of *immunological memory*. It had been noticed from the earliest days of immunology that exposure to an antigen resulted in a heightened state of responsiveness. The first time a given antigen comes into the system, the antibody response is rather sluggish, and antibody may not appear for several days. But all subsequent exposures to that same antigen result in a much quicker appearance of antibody (one to two days) and a much higher level of antibody in the blood. This came to be referred to as "memory" and is what we mean when we say we are "immune" to something. Instructionalist theories described very well the formation of antibodies, but it was known at the time that antibodies, once they clear antigen from the system, are themselves rapidly eliminated by breakdown and secretion through the urine. What then could explain the more rapid and higher-level antibody response in an animal exposed to the same antigen again two years later? Various speculations were put forth, but they were by and large unconvincing, and immunological memory remained unexplained.

However, the final blow to instructionalist theories would ultimately be delivered not by shortcomings in the theories themselves, but by newly emerging knowledge about how proteins—and thus antibodies—are made. By the mid-1950s, it was becoming increasingly clear that proteins cannot in fact alter their three-dimensional shapes simply as the result of contact with another molecule. Experiments in biochemistry, and in the newly emerging field of molecular biology, made it absolutely clear that the conformation of proteins is determined entirely by their amino acid sequence, which is in turn dictated entirely by the DNA sequence encoding them—*by their genes*. If there are thousands or even millions of different antibodies produced in an immunological lifetime, in response to that many different antigenic challenges, then it simply must be the case that animals come equipped with that many different genes for that many antibodies.

This realization took a while to seep in, because not everyone was equally prepared to understand and accept the newly emerging principles of molecular biology. And to accept such a theory would force immunologists to return to some version of Ehrlich's original theory, which had generally become anathema. But eventually the relation of protein structure to amino acid sequence, and ultimately DNA sequence, simply had to be accepted; the evidence coming in from protein chemists and from scientists studying DNA and genes was moving consistently and inexorably to precisely that conclusion. Suddenly it was as if the previous twenty-five years had simply not existed. The intuitive comforts of instructive theories were ripped away in a blink, with nothing left to replace them. Yet the Landsteiner refutation of Ehrlich's theory seemed as solid as ever. As with most transition periods, this was a time of genuine intellectual anguish and confusion for those who thought deeply about these problems.

In the end, these new understandings about proteins and DNA forced at least a few brave souls to reconsider the possibility, as counterintuitive as it might seem, that either the entire repertoire of antibodies necessary to deal with the antigenic universe, or at least the ability to generate such a repertoire, must be present in each of us at birth, as part of our normal genetic heritage. The first to take this leap in the modern era (1955) was the Danish immunologist Nils Jerne. Jerne found the new information about how proteins fold in three-dimensional space compelling, and he accepted the fact that the information for the primary structure of each protein is encoded in DNA. Because antibodies are proteins, and because there must be a large enough number of antibodies to deal with the entire antigenic universe, then it must be the case that each of us has a built-in set of genetic instructions (genes) to make all possible antibodies for all possible antigens. Examples of these antibodies must exist prior to contact with any given antigen, and antigen must somehow select out from this built-in repertoire just those antibodies needed to deal with any given incoming antigen.

Jerne proposed that samples of all the various antibodies an organism is prepared to make would be present at all times, albeit in low levels, in the blood. When antigen entered the system, and combined with one or more of these antibodies, those antibodies thus "selected" by antigen would be transported to some unspecified cell where they would be reproduced in great quantities and released back into the circulation. Without referring specifically to Ehrlich, Jerne took refuge in Darwin and called his theory the "natural selection theory of antibody formation." His paper stirred an immediate and vigorous debate. As can be imagined, it was roundly criticized by the proponents of instructionalist theories. But Jerne was clearly in harmony with the rapidly advancing understanding of protein structure, a point even the most diehard instructionalists had to concede. On the other hand, his proposal that antibodies could somehow serve as templates for the production of more copies of themselves went in the opposite direction from molecular biology and presented a weak point his critics were quick to attack. All the emerging laboratory evidence suggested that proteins are synthesized using a messenger RNA (mRNA) copy of the gene for that protein. Jerne was proposing that somehow a protein (the antibody) could be directly copied into another protein, with no intermediate messenger of any kind. There was simply no experimental support for such a proposal.

Without abandoning the basic tenet of antigen selection of the antibody to be produced, first David Talmadge, and then almost immediately (and perhaps more forcefully) the Australian immunologist F. Macfarlane Burnet, proposed that antigen does not select individual antibody molecules per se for reproduction by some unknown copying mechanism. He proposed, rather, that antigen somehow selects individual *cells* capable of making individual antibodies. Burnet had been thinking about this problem for many years, and he had previously put forward the only serious alternative to instructionalist theories. He was convinced that antibody-forming cells must play a key role in the immune response, and he had criticized instructionalist theories as the creation of chemists who ignored the cellular basis of the synthesis of biological macromolecules.

Burnet was very vigorous in his promotion of this idea and fielded criticisms of it effectively, rapidly modifying his theory's structure when necessary, even incorporating some of his critics' ideas, where appropriate. Burnet spoke of individual antibody-forming cells, each committed to the production of one and only one antibody. He proposed that such cells would display a sample of that antibody on their surface, much like a merchant displays a sign on his shop to let people know what it is he sells. When the cell encountered an antigen that it could bind through this antibody, the cell would be stimulated to do two things: commence production and secretion of antibody, and begin to divide, giving rise to clonal progeny all capable of making the same antibody. Burnet called this the *clonal selection theory* of antibody production. His theory was overpowering in its simplicity, and in its ability to account for all of the features of the immune system known at the time. Although challenged vigorously for the next several years, it has survived to become essentially *the* central dogma of all of immunology. Burnet received the Nobel Prize for his work in 1960. The Clonal Selection Theory posits the following:

1. There exists a specific subset of cells in each animal intended solely for antibody production. Antibody production is not a side product of some other biological function such as nutrient uptake.

2. The central units of this system, the antibody-producing cells, each are dedicated to the production of one and only one antibody. Each cell is capable of producing a different antibody, with different antigen-combining-sites; it displays a copy of that antibody on its surface. This antibody is generated randomly, without reference to or knowledge of the antigenic universe. If a given cell is stimulated through its surface antibody, events leading to production of that antibody will be initiated.

3. In addition to stimulating initiation of antibody production, binding of antigen to the surface antibody will also stimulate the cell to divide, giving rise to large numbers of clonal progeny all dedicated to producing the same antibody. These cells (or their progeny) live on after the initial round of antibody production.

4. Any clones produced through this mechanism that cross-react with self molecules will be eliminated so as to avoid problems of autoimmunity.

This simple hypothesis accounts for the three central features of the immune system, as follows.

1. *Diversity.* This is what led us into this historical discussion in the first place. The immune system cannot compete with

rapidly evolving microbes by genetically tailoring its response to the characteristics of each potential pathogen on an ad hoc basis. The pathogens can mutate in response to the immune system's adaptations too rapidly. Therefore, the immune system evolved the ability to generate enormous antibody diversity capable of recognizing any conceivable pathogen over all of evolutionary time. F. Macfarlane Burnet did not offer a mechanism for doing this in his original proposal. This was of course focused on by the critics of clonal selection. Even after clonal selection became generally accepted, the question of the "generation of diversity" (GOD, as it came fondly to be called) would remain a central question in immunology for many years. Exactly how the immune system generates this diversity we shall explore in a moment. For now let us simply accept that *the ability to generate a complete antibody repertoire is available in each of us at birth*.

2. *Tolerance of self.* In trying to look at the world from the immune system's point of view, we might ask whether there are any limitations placed on the immune system as it carries out its central task. Clearly, there is one major restriction. Whatever mechanism is used to generate immune responses, they must never be directed at the host organism itself. Thus, in generating the diversity of recognition necessary to identify and destroy all possible pathogenic invaders, we must either not generate anti-self reactivity or, if we do, we must either destroy it or keep it tightly regulated. Nils Jerne had proposed that antibodies reactive with self would bind to self tissue, and thus not be available for transport into cells and large-scale reproduction. Burnet recognized the importance of this problem as well, and the weakness of Jerne's solution. He proposed that if clones of antibody-producing *cells* reactive with self ("forbidden clones") were in fact generated, they would have to eliminated. As with the generation of diversity, he was vague on this point, and was again attacked for it. It was probably just

as well he was vague, for this turns out to be one of the trickiest questions in immunology. In fact, many anti-self clones are *not* eliminated, but rather kept under control. This contains the seed of a problem, in the form of latent autoimmune disease.

3. *Memory.* As pointed out previously, the inability of instructionalist theories to provide a satisfactory explanation of immunological memory was another of their great failings. Even the instructionalists had to concede that Burnet's explanation of memory was powerfully attractive. The result of initial exposure to antigen is a greatly expanded and stabilized set of cells capable of producing a specific antibody. When the same antigen reenters the system, the number of antibody-producing cells that recognize it has expanded perhaps ten thousand–fold, and this is reflected in the vigor of the so-called *secondary* or *anamnestic* (recall) response. This is a major feature distinguishing the vertebrate immune system from the defense systems of lower animals. The latter certainly do have defenses against invasion by pathogens, but these defenses are not *adaptive,* that is, they are not changed as a result of *experience,* of contact with antigen. Such defense systems have no memory of things past.

Although accepted almost immediately by a handful of immunologists, clonal selection by no means swept the field at its debut. Here is a statement taken from a prominent textbook published in 1966, nine years after Talmadge and Burnet proposed clonal selection as an alternative to instruction. It was written by William Boyd, an outstanding immunologist of the time:

Like Haurowitz [a leading proponent of instructionalist theories] I find it "difficult to believe that the body should contain preformed antibodies against azophenylarsonate, azophenyltrimethylammonium ions, and other artifacts of the chemical laboratory." So in spite of the present vogue of the "selection" theories, I incline to believe that they will ultimately prove invalid.

The value of any theory is ultimately judged not by its ability to explain, but by its ability to predict. The clonal selection theory of antibody production by that criterion proved valuable indeed. It predicted the existence of a cell, dedicated to antibody production, that would have a copy of the antibody it was prepared to produce posted at the surface. After a good deal of searching, such a cell was indeed found: the B lymphocyte, or B cell. Clonal selection predicted that this cell would produce one and only one kind of antibody. Careful experimentation established this fact as well. Both the predicted clonal expansion of B cells and their involvement in memory have been demonstrated. Clonal selection also predicted the elimination of self-reactive antibody-forming cells; in the past several years we have seen this prediction also bear fruit. Although under constant scrutiny for flaws, clonal selection is nevertheless still the set of rules by which almost all immunological phenomena are interpreted, and upon which almost all immunological experiments are planned. Past history tells us only too well that scientific dogma rarely survives intact more than a generation or two; but at present, clonal selection still does an invaluable job in helping us to both explain and predict immunological phenomena.

Immunology after clonal selection, like the newly emerging field of molecular biology, seemed to move forward with a steady force that gradually swept everyone along with it. A number of developments in the early 1960s helped make clonal selection seem more rational and less counterintuitive and ultimately convinced even the most diehard instructionalists, such as Linus Pauling himself. I will cite only two examples here. The first was an experiment published in 1964 by Haber, which is generally conceded to be the most direct proof of the molecular biologists' contention that the folding of antibodies in three-dimensional space, and thus the creation of their antigen-combining sites, must occur independently of antigen. Haber took a sample of antibody that had been produced against a specific antigen and subjected it to chemical conditions that caused it to *denature*— to unfold, to completely lose its three-dimensional structure. In this condition, the antibody protein was just a formless string of amino acids (the building blocks of proteins) with no ability to bind antigen to the instructionalists, it was just a blank template. Haber must have held his breath as he then reversed the denaturing reaction, slowly readjusting the antibody's chemical environment to physiological conditions, *in the absence of antigen*. He then added this *renatured* antibody to a sample of antigen. It recognized the antigen instantly and bound to it, but to *no other antigen*. It apparently had been able to fold itself from a shapeless string of amino acids into a fully functional antibody molecule *without any influence of antigen*, just as the molecular biologists had predicted. The proposed dependence of antibodies on antigen for creation of their antigen binding sites was at the core of instructionalist theories. Few diehards remained after this experiment was published.

The second development that helped rationalize clonal selection for many scientists is what I call "the taming of the immunological universe." The experiments of Karl Landsteiner had suggested that the antigenic universe must be quite large. In a sense, it seemed infinite. Investigators kept finding that basically any molecule they made, as long as they made it from those atoms commonly used to make biological macromolecules, would elicit antibody formation when injected into an animal. But therein lay a key fact, one that would ultimately place an important restriction on the size of the antigenic universe: the number of different atoms from which antigenic molecules could be constructed. All living things on this earth—all potential antigens and all potential pathogens—are made from only half a dozen or so atoms: carbon, hydrogen, oxygen, nitrogen, sulfur, and phosphorus.

There was another fact that helped limit the size of the antigenic universe. The immunochemists had been saying for years that antibodies would almost certainly have discrete sites somewhere in their structure that would be involved in antigen binding. Experiments by William Boyd himself in the 1940s had already suggested the number of such sites would be two per antibody molecule. Now the *size* of these sites became of paramount importance, because that would predict the size of the antigen that could be bound. Within a short time this was determined to a first approximation by Elvin Kabat. The portion of an antigen molecule that would fit into an antigen-combining site on an antibody molecule was somewhere on the order of four or five sugar molecules, or five to six amino acids.*

This was a major breakthrough. Now the size of the antigenic universe could be estimated. The portion of an antigenic molecule "seen" by an antibody, which we call an *antigenic determinant,* was on the size order of, say, five amino acids. Amino acids, like all biological molecules, are composed of only six or so different atoms. We know the average number of such atoms used in making an amino acid. Given this size restriction, how many antigenic determinants could one possible make from only six atoms? This is the period of time (mid-1960s) when I came into immunology myself. As graduate students we sat around playing with numbers, trying to predict the size of the antigenic universe. Like others, our numbers always came out somewhere around 10^8 to 10^{10}, that is, somewhere between a hundred million and ten billion possible different antigens. Large numbers, indeed, but far from infinite. The next game was to try to predict how much DNA would be required to encode that many antibody molecules. Would there even be enough DNA in the entire human genome? Depending on one's assumptions, the figures varied from less than 1 percent of the human genome up to as much as 10 percent.

---

*This is an important point to grasp. When we refer to something like a bacterium or a virus as an antigen, from the immune system's point of view it is really a collection of antigenic determinants. There is a huge difference in scale: An antibody is a protein, millions of times smaller than any living cell. When an antibody binds to the surface of a bacterium, it is binding to an incredibly tiny portion of its surface—a single antigenic determinant thereon. It also follows from this that a bacterium may induce the production of a very large number of different antibody molecules, depending on how many different antigenic determinants it displays on its surface.

As it turns out, both figures are gross overestimations. In science, guessing games are only played in the absence of hard experimental data. Beginning in the late 1970s, the tools of molecular biology made it possible to simply go in and have a look at antibody genes in an animal's DNA, count them, and see how they work. What the molecular biologists found was astonishing and completely unexpected; it revolutionized how we think of genes and proteins. It also finally made clear how vertebrates are able to make the virtually unlimited numbers of different antibodies necessary to deal with a large and rapidly changing antigenic universe.

It is the B cell, remember, that is charged with making antibody. Given that the B cell must be able to make a hundred million or more different antibodies, it could be imagined that B cells as a group must come equipped with a hundred million or more different genes from which to make them. That was viewed as an upper limit, but was actually the opinion held for many years by one school of thought on the GOD, or generation of diversity, question. The solution to the diversity problem in this case would be to equip each B cell with an extremely large—but fixed—number of antibody genes, from which one and only one would be expressed.

But the immune system turns out to be much more clever than that—*much* more clever. Rather than make antibodies from a fixed, inherited pool of antibody genes, the immune system allows each B cell to assemble its own antibody genes de novo—from scratch! Antibody genes per se are not inherited at all. Each B cell inherits several pools of gene fragments, sort of like several bags of different-shaped Lego pieces, from which it assembles antibody proteins. This is absolutely unique in all of biology. In no other system (as far as we know) are the genes used to build proteins assembled within the individual, rather than inherited from one's parents.

An example of how this works is shown in Figure A.1. One of the protein chains used to make an antibody molecule is the so-called H chain. The gene for the part of the H chain that is important in binding to antigen (and which thus has to be *different* for each antibody molecule) is actually constructed from three different pools of gene fragments inherited by each B cell. The average B cell will have about 300 different fragments in the V pool, 12 or so in the D pool, and about 6 in the J pool. Now, just on the basis of random conbination alone, a B cell could use these 320 or so fragments to assemble 300 $\times$ 12 $\times$ 6, or over 20,000 different antibody chains. But wait! The immune system is even more clever than that. It allows the B cell to assemble these three segments *imprecisely*. That is, a V segment does not have to be joined perfectly end-to-end with a D segment. It can overlap the D segment by varying degrees within fairly wide limits. Or the V segment may be brought up short of the D segment, and the intervening space can be filled in with random DNA material. The same is true for the junction between the D segment and the J segment. Each of these various gene assemblages is different and codes for a unique antibody molecule, each capable of binding to a different antigen.

So from the originally inherited 300 or so small pieces of DNA, the B cell can assemble over 20,000 unique antibody H-chain genes by randomly (but perfectly) matching these fragments; by allowing imprecision of joining, the number of different antibody proteins that can be created is enormous. Accurate estimates are impossible, but the number certainly far exceeds even the wildest estimates of the variability needed to deal with the antigenic universe.

It would be difficult to imagine a more elegant solution of the problem posed to the immune system: Devise a way to deal with a universe of pathogens that is not only enormous to begin with but full of pathogens that can alter themselves genetically hundreds of thousands of times faster than vertebrates can. The answer: Bypass standard methods of shuffling the genetic deck through normal breeding processes. Come up with a whole new system for mutating genes that allows you to create not hundreds of *thousands* of

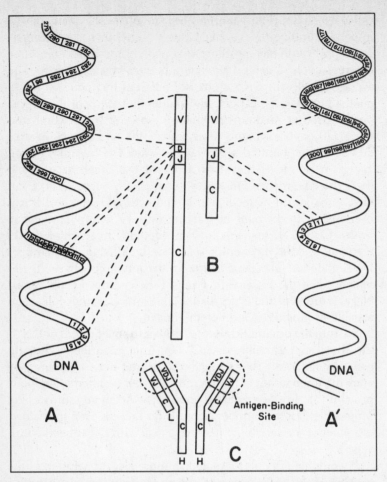

FIGURE A.1. *Assembly of an antibody molecule.* As shown at the bottom of the figure, a complete antibody molecule consists of four protein chains: two heavy (H) chains, and two light (L) chains. The part of the antibody molecule that binds to an antigen is located at one end of each pair of H and L chains. This "antigen-binding site" is identical for both halves of the same antibody molecule but is different between different antibodies. It is this end of the molecule that is responsible for dealing with the antigenic universe. As shown in (**B**), the portion of each chain that is involved in antigen binding consists of three distinct parts (V, D, and J) for

new antibodies per *generation*, but hundreds of *millions* of new antibody molecules per *hour!* Every hour. All life long. Let a pathogen try to get around *that!*

So, in the end, Paul Ehrlich had it right—*and* he was first. Antibodies do preexist in the body, independent of any knowledge about antigen. The B cells simply sit there, cranking out millions upon millions of different antibodies every day, creating a huge antibody repertoire that constantly shifts and changes, faster than pathogens can mutate to escape it. But Ehrlich also forsaw the possibility of treason and tragedy in such a system, and again he was right (and first). During this rapid-fire assembly of antibody molecules with completely random antigen specificity, how do we prevent them from assembling antibodies that react with *self* antigen? How do we make sure that the full fury of the immune system stays trained on outside invaders, and is not used against us? Burnet tried to address this problem in his clonal selection theory, as we just saw; mostly he just waved his hands. Ehrlich in his day referred to this possibility as *horror autotoxicus*. This may seem like a rather dramatic name for what nearly everyone of his time regarded as an impossibility. As happened so very often, Ehrlich was right and the others were wrong. But I'm not sure even Ehrlich could have imagined how wrong they might be!

FIGURE A.1. (*Continued*) H chains, and two parts (V and J) for L chains. Each of these parts is encoded in the DNA as a separate gene segment. For an H chain, any V-gene segment can be combined with any D segment, which can be combined with any J segment, to produce the final antigen-combining region. Similarly, an L-chain antigen-combining region is made by random combination of a V- and a J-gene segment (L chains do not use D segments). The C regions of each antibody molecule do not vary much between different molecules. They are encoded by a separate set of genes and serve largely to hold the antigen-combining sites in the proper orientation.

Bibliography

Books and Monographs

Bibel, Debra J. *Milestones in Immunology*. Science Tech Publishers, Madison, Wisc., 1988.

Blalock, J. Edwin (Ed.). *Neuroimmunoendocrinology*, 2nd ed. Chemical Immunology, vol. 43. S. Karger, Basel, 1992.

Blumstein, J., and F. Sloan. *Organ Transplantation Policy: Issues and Perspectives*, Duke University Press, Durham, N.C., 1989.

Dixon, C. W. *Smallpox*. J. A. Churchill, London 1962.

Fisk, Dorothy. *Dr. Jenner of Berkeley*. W. H. Heinemann, London, 1959.

Land, W., and J. B. Dosseter (Eds.). *Organ Replacement Therapy: Ethics, Justice, Commerce*. Springer-Verlag, Berlin, 1991.

Lloyd, Ruth. *Explorations in Psychoneuroimmunolgy*. Grune & Stratton, Orlando, Fla., 1987.

Marquardt, Martha. *Paul Ehrlich*. Henry Scuman, New York, 1951.

Metcalfe, Dean D., Hugh A. Sampson, and Ronald A. Simon. *Food Allergy: Adverse Reactions to Foods and Food Additives*. Blackwell Scientific Publications, Boston, 1991.

Samter, Max, David Talmadge, Michael Frank, K. Frank Austen, and Henry Claman (Eds.). *Immunological Diseases*, 4th ed. Little, Brown, Boston, 1988.

Silverstein, Arthur M. *A History of Immunology*. Academic Press, New York, 1989.

Stiehm, E. Richard (Ed.). *Immunologic Disorders in Infants and Children*, 3d ed. W.B. Saunders, Philadelphia, 1989.

Stine, Gerald J. *Acquired Immune Deficiency Syndrome: Biological, Medical, Social and Legal Issues.*. Prentice Hall, Englewood Cliffs, N.J., 1993.

Thucydides. *The Peloponnesian War.* Translated by Rex Warner. Penguin Classics, Harmondsworth, 1972.

Turk, J. L. *Delayed Hypersensitivity.* Elsevier/North Holland Press, Amsterdam, 1980.

Valery-Radot, D. *The Life of Pasteur.* Doubleday, Page & Co., New York, 1927.

Wagner, Richard. *Clemens von Pirquet: His Life and Work.* Johns Hopkins University Press, Baltimore, 1968.

Original Articles

Ballieux, R. E. 1992. Bidirectional communication between the brain and the immune system. *European Journal of Clinical Investigation* 22 (Suppl. 1):6.

Blalock, J. Edwin. 1994. The immune system: Our sixth sense. *The Immunologist* 2:8.

Bovbjerg, D., W. Redd, L. Maier, J. Holland, L. Lesko, D. Niedzwicki, S. Rubin, and T. Hakes. 1990. *Journal of Consulting and Clinical Psychology* 58:153.

Bruton, Ogden C. 1952. Agammaglobulinemia. *Pediatrics* 9:722.

Burnet, F. M. 1957. A modification of Jerne's theory of antibody production using the concept of clonal selection. *Australian Journal of Science* 20:67.

Cohen, S., D. Tyrrell, and A. Smith. 1991. Psychological stress and susceptibility to the common cold. *New England Journal of Medicine* 325:606.

DiGeorge, A. M. 1965. A new concept of the basis of cellular immunity. *Journal of Pediatrics* 67:907.

Dropulić, B., and K. Jeang. 1994. Gene therapy for human immunodeficiency virus. *Human Gene Therapy* 5:927.

Goetzl, E., and S. Sreedharan. 1992. Mediators of communication and adaptation in the neuroendocrine and immune systems. *FASEB Journal* 6:2646.

Haber, Edgar. 1964. Recovery of antigenic specificity after denaturation and complete reduction of antibody. *Proceedings of the National Academy of Sciences* (USA) 52:1099

Hurwitz, Samuel H. 1929. Hay fever: A sketch of its early history. *Journal of Allergy* 1:245.

Jerne, Nils. 1955. The natural selection theory of antibody production. *Proceedings of the National Academy of Sciences (USA)* 41:849.

Kiecolt-Glaser, J. K., and R. Glaser, 1991. Psychosocial factors, stress, disease and immunity. In R. Ader, D. Felteo, and N. Cohen (Eds.), *Psychoneuroimmunology*. Academic Press, New York, 1991.

May, Charles D. 1985. The ancestry of allergy: An account of the original experimental induction of hypersensitivity recognizing the contribution of Paul Portier. *Journal of Allergy and Clinical Immunology* 75:485.

Merrill, John P., J. E. Murray, and W. G. Guild. 1956. Successful homotransplantation of the human kidney between identical twins. *Journal of the American Medical Association* 160:277.

Rogers, M., and P. Reich. 1988. On the health consequences of bereavement. *New England Journal of Medicine* 319:510.

Silverstein, A. M. 1979. History of immunology. Cellular vs. humoral immunity: Determinants and consequences of an epic 19th-century battle. *Cellular Immunology* 48:208.

Solomon, George F., and Rudolph Moos. 1964. Emotions, immunity and disease: A speculative theoretical integration. *Archives of General Psychiatry* 11:657.

Thrusfield, Hugh. 1941. Smallpox in the American War of Independence. *Annals of Medical History* (Ser. 3) 2:312.

Weigent, D., D. J. Carr, and J. Edwin Blalock. 1990. Bidirectional communication between the neuroendocrine and immune systems. *Annals of the N.Y. Academy of Sciences* 579:17.

Index

causative agent of, 7, 26
description of, 7
eradication of, 6, 25
inoculation, 8
relation to cowpox, 23
Solomon, George, 225
Speciesism, 211
Specificity, immunological, 40
Starzl, Thomas, 208
Stem cells, 51, 75, 119, 135
Stress, 95, 223, 227
Suicide, cellular, 43
Systemic lupus erythematosis, 123, 126–27

Tagliacozzi, Gasparo, 179
Talmadge, David, 255
TB. *See* Tuberculosis
T cells, 38, 41–44, 54
helper (CD4), 43, 44, 49, 55; in AIDS, 154–60
killer (CD8), 43, 50, 116; in AIDS, 157, 166; in graft rejection, 185; in tuberculosis, 111–17
Tetanus, 36
Thrush, 143
Thucydides, 4, 221
Thymus, 53–55, 63
in myasthenia gravis, 129
Thyroglobulin, 121

Thyroid stimulating hormone (TSH), 223
Thyroxin, 121
Tolerance, immunological, 40, 117–20, 252, 257
Tuberculin, 110
Tuberculosis, 31, 108–14
in AIDS patients, 145
causative agent of, 109

Uniform Anatomical Gift Act, 199

Vaccination, 19–26
in SCID infants, 67
Vaccinia virus, 19
Variola virus, 7
Variolation, 22
Vasoactive intestinal peptide (VIP), 232
Virulence, 112
Viruses, 50, 114
in AIDS, 144

Wagstaffe, William, 14
Willis, Thomas, 127

Xenotransplantation, 205
X-linked diseases, 62, 68

Zidovudine. *See* Azidothymidine

Pursuant of the Uniform Anatomical Gift Act.
I hereby elect upon my death the following option(s):

A ___ To donate any organ or parts.
B ___ To donate a pacemaker (date implanted _____).
C ___ To donate parts or organs listed _____
_____.
D ___ To not donate any organs, parts or pacemaker.

_____
    Signature   Date

A valid facsimile of a Uniform Anatomical Gift Act donor card. It will be
recognized and honored by any medical center in the United States.
Simply cut it out and paste or tape it to the back of a current valid
driver's license or other form of photo ID.